T0279663

The Honey Trap

The Honey Trap

How the Good Intentions
of Urban Beekeepers
Risk Ecological Disaster

DANA L. CHURCH

sh.

SUTHERLAND
HOUSE

TORONTO, 2024

Sutherland House
416 Moore Ave., Suite 304
Toronto, ON M4G 1C9

Copyright © 2024 by Dana L. Church

All rights reserved, including the right to reproduce this book or portions thereof in any form whatsoever. For information on rights and permissions or to request a special discount for bulk purchases, please contact Sutherland House at sutherlandhousebooks@gmail.com.

Sutherland House and logo are registered trademarks of The Sutherland House Inc.

First edition, September 2024

If you are interested in inviting one of our authors to a live event or media appearance, please contact sranasinghe@sutherlandhousebooks.com and visit our website at sutherlandhousebooks.com for more information.

We acknowledge the support of the Government of Canada.

Manufactured in Turkey
Cover designed by Leah Ciani, and Jordan Lunn
Book composed by Karl Hunt

Library and Archives Canada Cataloguing in Publication
Title: The honey trap : how the good intentions of urban beekeepers risk ecological disaster / Dana L. Church.
Names: Church, Dana L., author.
Description: Includes bibliographical references.
Identifiers: Canadiana (print) 20240389263 | Canadiana (ebook) 20240389298 | ISBN 9781990823855 (softcover) | ISBN 9781990823862 (EPUB)
Subjects: LCSH: Urban bee culture—Environmental aspects. | LCSH: Bees—Effect of human beings on. | LCSH: Honeybee.
Classification: LCC SF523 .C48 2024 | DDC 638.109173/2—dc23

ISBN 978-1-990823-85-5
eBook 978-1-990823-86-2

For the bees

Contents

Introduction

A FEW YEARS AGO, I began noticing the explosion in urban beekeeping across Canada, the United States, and Europe. Beehives were popping up in more and more cities, and the number of hives in those cities kept creeping higher and higher. Large North American turnkey urban beekeeping companies—companies that rent out and manage beehives for corporate, non-profit, or residential clients—were promoting urban beekeeping as a path toward "greener" cities. As a trained scientist with an interest in bees, particularly bumble bees, I began to feel this was a bad idea. A very bad idea. And I wasn't alone. Scientists had been studying this and issuing warnings, but no one seemed to be listening.

There have been glimmers of hope. In May 2023, the *Toronto Star* published an article titled "Why I'm Giving Up Beekeeping: Urban Beekeeping Is, Paradoxically, Bad for Bees." The author was Jode Roberts, one of a number of urban beekeepers in Toronto, where he had maintained honey bee hives for years. He had realized that as good as his intentions were, his hives were part of a growing problem.

Appearing in the mainstream media where it would reach thousands of people, Roberts's article came as a relief. Here, at last, was someone writing for the broader general public making the point that, whatever else we felt, urban beekeeping was "bad for bees."

This may seem paradoxical. We're talking bees here. Honey bees have been adored for centuries. They give us honey, and they pollinate flowers, including our food crops. How can urban beekeeping be a bad thing?

Because there are a number of misconceptions about honey bees. I believe two directly contribute to the popularity of urban beekeeping: honey bees are going extinct; and honey bees are "good," so more is better.

Despite the devastating losses beekeepers continue to experience in some regions of the world, honey bees are not going extinct. For instance, queen honey bees—each of which can lay enough eggs to produce a colony of 50,000 or more honey bees—can be ordered over the internet. And second, when it comes to pollination, depending on the region and plants in question, managed honey bees can *worsen* fruit quality and yields. I've heard a number of bee scientists comment that keeping honey bees to save the bees is like keeping chickens to save the birds.

The honey bee that most people are familiar with, the western or European honey bee, *Apis mellifera,* is only one of approximately 20,000 different species of bees on the planet. There are stingless bees, like *Melipona beecheii* found in Central America, specialized bees such as the squash bee which relies on the flowers of that plant, highly social bees, such as the honey bee, and other bees that are completely solitary creatures. There's also my particular area of research, the bumble bee.

In this book, I showcase a number of non-honey bee species to provide a glimpse of this diversity. I also show that our relationship with honey bees goes back to when humans first discovered honey bees, and more important, honey does not exist in a vacuum. How we treat and use honey bees can and does have far-reaching and potentially devastating consequences for other bees. There can also be knock-on effects on plants, both cultivated and wild, and ecosystems more generally. (Full disclosure, another reason I showcase different bee species is because they are incredible! There is so much to love about bees besides honey. One of my intentions is to spread the wonder, love, and admiration that I have for all types of bees, big or small, furry or hairless, because they do so much for the world and have so much to teach us.)

In what follows, I have tried my best to weave a straightforward story. But when it comes to modern times, the story of humans and bees has exploded into something tangled and nuanced. We have thrown so many things at bees: habitat fragmentation, pesticides, crop monocultures, and climate change to name a few. The story spreads in so many different directions that it kind of reminds me of the string art my dad made in the 1970s: It had multitudes of threads of various colors stretched all over the felt-backed canvas. Baffling, but if you stepped back, you could see that taken as a whole, the strings formed a complete picture. (In my dad's case, it was a sailboat.)

I hope that by the end of this book you can see a clear picture of how misguided intentions, assumptions, and ideas about urban beekeeping are fuelling its spread, and how the expansion of urban beekeeping shows how dangerously narrow and selfish our thinking can be. The expansion of urban beekeeping is the latest, and arguably the most dangerous, indignity we have visited upon the bees. We are paving the road to ecological disaster with our good intentions.

Let's begin.

Chapter One

L ET'S GO BACK in time, more than 8,000 years. We are near Valencia, Spain. The summer sun shines in a brilliant blue sky. The air is fresh, the rivers run clear, and the world teems with animal and insect life. There are no man-made buildings or structures anywhere; just an endless landscape of natural beauty. Up ahead we can see a tall rock cliff. Suspended up high at an alarming height is a naked man, dangling from a flimsy-looking rope. In one hand he is holding what looks like a basket, and with the other, he is reaching inside a crevice in the cliff wall. We can see bees swarming around his whole body, and if we listen, we can hear them buzzing. This man has risked his life climbing so precariously high, and he will likely suffer a number of bee stings. All in order to ransack this honey bee nest.

The man is stealing from *Apis mellifera*, the European honey bee (also called the western honey bee). He is scooping out wax comb. Most of it is filled with honey, but some of the comb contains eggs, or larvae, or pupae—the three separate stages of a developing honey bee—as well as stored pollen. (A larva is the worm-like creature that hatches from the egg. As it grows, it changes into a pupa, the stage when it looks more and more like a bee.) All are good sources of protein. When his basket is full, he will climb back down, return to his family, and share his booty; but perhaps he will sneak a taste or two first.

If we can imagine what life must have been like for this man and his family thousands of years ago, finding a honey bee nest would have been like discovering hidden treasure. Quite literally. At the time, *A. mellifera* were not the same bees that we keep in purpose-built hives today. Instead, they were wild bees that built their nests inside trees, logs, rock walls, anywhere that was hidden and provided protection from the elements. Humans, such as the brave, rock-climbing man, likely found honey bee nests while hunting and gathering. Perhaps they witnessed another animal pillaging a nest and wondered what prize they had found, or perhaps they simply saw buzzing honey bees around a nest entrance and smelled the heady scent of honey. Curious, they likely poked a stick into the nest, saw liquid gold dripping from the stick when they pulled it out, and ventured a taste.

In a world where your diet consisted of whatever meat, berries, or fruits you can find, that first taste of honey must have been mind-blowing. Nothing else would have tasted so sweet. No wonder that after experiencing a taste of honey, people started seeking out honey bee nests and stealing from them when they could, even if it meant risking your life by climbing great heights.

Our friend wouldn't have known it, but there was more to honey than a great taste. Made up of 80 to 90 percent sugar, it is one of the most energy-dense foods found in nature. Liquid honey is a concentrated source of fructose and glucose and contains trace amounts of several essential vitamins and minerals. The honey that Paleolithic people ate would also have contained trace amounts of bee larvae, which is a good source of protein, fat, several essential minerals, and B-vitamins. (People might have specifically harvested bee larvae from bees' nests in addition to honey.) Honey would have been an excellent supplement to the meat and plant diet of early humans. Recently, some scientists have suggested that honey has been overlooked as an important component of the early human diet. It would have provided the extra calories needed to help fuel the growth and evolution of the human brain, and glucose is crucial for neural development and function. With tools that would have allowed them to harvest honey more easily and efficiently, early humans would have had a nutritional advantage over other species. Perhaps we should be thanking bees at least in part for our evolutionary success.

Today if we look on the walls inside what is known as La Araña—the Cave of the Spider—near Valencia, Spain, we will see that naked man's honey-hunting documented in an ancient painting. It shows a human figure, surrounded by bees, who has climbed what appears to be a rope ladder to raid a bees' nest high above the ground. Whether the honey-hunter himself painted this or it was done by a witness, the act of gathering honey made such an impression that someone thousands of years ago decided to record it. What was going through the minds of those ancient peoples and the artist who commemorated the gift of honey? How must they have perceived the bees? Did they worship them? Did they use the painting as some kind of altar? Or was the painting simply an expression, or a reminder, of the impact those tiny, buzzing insects had on people whose lives revolved around finding enough food? In any event, that painting in the Cave of the Spider is regarded as among the earliest pieces of evidence of the relationship between humans and honey bees.

For thousands of years, stingless bees have held great cultural significance for the Indigenous peoples of Africa, Asia, South America, and Australia. Their honey has been used for food as well as medicine, and their wax to make art, tools, weapons, and gifts. These bees feature extensively in traditional stories, song, and dance. The Mayan people in particular developed a deep relationship with stingless bees, worshipping them as a gift from the gods, and considering them a sacred animal. At some point, people began "keeping" bees, so they could more easily reap the benefits of honey and wax. There is no clear indication as to who were the very first "beekeepers," but ancient Mayans kept native stingless bees in hollow logs, called *jobones*, that were about two to four feet long. The ends were plugged with clay or stone discs to keep out ants and other predators. An entrance hole for the bees, often marked with a cross, was made in the middle of the log. To harvest the honey, the beekeeper simply had to remove the plug at one end and reach inside. These jobones were stacked on an A-shaped frame and kept shaded within a thatched hut called the *nahil-cab* or bee house. Archaeologists have found many of the stone discs that were used as plugs for jobones, all elaborately carved with Mayan glyphs. Other Mayan artifacts depict deities associated with beekeeping and honey harvesting. As well, a Mayan codex currently displayed in Madrid (a codex is a folding book made on paper produced from tree bark), shows hieroglyphs and images of gods, rituals, and activities related to the keeping of beehives and harvesting of honey.

Beekeeping in India can be traced back to around four thousand years ago, when people began making hives for native species of honey bees out of clay pots or twigs and grasses covered in dried mud. These were often hung from the ceiling in farmyard structures or kept in wall niches. Interestingly, ancient Indian beekeepers used a number of practices that have stood the test of time. For instance, like modern beekeepers, they used smoke to calm the bees while they harvested the honey, and they left enough honey behind to ensure that the bees remained well fed and healthy. If a hive became overpopulated, they split the colony and created a new one. Like in other cultures, the importance of bees was reflected in Indian rituals. For example, brides were anointed with honey to ensure fertility, and honey was served to guests after the marriage ceremony to ward off any evil spirits and ensure that the couple would have a happy and prosperous life.

The ancient Egyptians were some of the earliest beekeepers, as evidenced by many cave drawings and temple paintings dating back to at least 2400 BCE that illustrate the harvesting, processing, and storing of honey. The Egyptians worshipped bees and thought they came from tears shed by the sun god, Ra. Ancient Egyptian beehives were wicker baskets covered in clay and baked in the sun until hard. Sometimes these hives were placed on rafts and floated down the Nile, so that the bees could visit the different riverside flowers and thus produce unique blends of honey. Honey produced from temple apiaries was believed to have special healing properties, so it was used to make medicines and ointments. (An apiary is a location where honey bee colonies are kept and can vary in size.) Ancient Egyptians also harvested beeswax, which they used for mummification and shipbuilding, and as a kind of gel to keep their elaborate wigs slicked down. Honey bees were so important to the Ancient Egyptians that the honey bee hieroglyph was chosen to represent the entire region of Lower Egypt. Elsewhere in the ancient Mediterranean world, the Greeks and Romans were also avid beekeepers.

In Medieval times, forest beekeeping was popular in northern Europe and also in Russia. Upon finding a honey bee nest inside a tree, a beekeeper carved a symbol into the bark, staking claim to both the tree and the bees. They often enlarged the nest entrance in the tree to make honey and wax harvesting easier. If the nest was high up, they cut footholds into the trunk to make climbing easier. In Russia, these beekeepers were called *bortniks*, where *bort* means "hollow tree trunk." There were laws to protect the bees, such that anyone who destroyed a bee tree was slapped with a heavy penalty. A number of woodcut prints in medieval manuscripts illustrate scenes of forest beekeeping.

In other parts of Medieval Europe, people kept colonies of honey bees in bell-shaped "skep hives" woven from straw coils. These are the traditional round beehives that even today are used to represent bees, for example on the state flag of Utah. Cow dung smeared on the outside acted as waterproofing. More straw coils, called "ekes," could be added to the bottom of the skep hive if the bees needed more room as the colony grew. When it was time to harvest the honey inside, beekeepers would often suspend the skep hive over a fire pit so the smoke would calm the bees. Occasionally, this could result in the skep hive bursting into flames!

Overall, what became clear to many people around the world was that honey bees could be managed. It was possible to control where they lived to simplify honey harvesting. Easy access to their wax was extremely important too, especially before the invention of electricity. Beeswax candles burn much longer and brighter than candles made of tallow (animal fat), and they smell much more pleasant. Another reason for humans to love honey bees.

As if a seemingly endless supply of honey and wax wasn't enough, bees provided human culture with an additional form of magic: mead. Some argue that mead, also called "honey wine," is the oldest alcoholic beverage, predating wine made from grapes and beer made from barley. Mead is made by mixing honey with a bit of water and leaving the mixture to ferment. All it probably took was prehistoric honey hunters leaving some harvested honey out in a rainstorm in warm weather. When they tasted it later, they would discover the curious sensation of tipsiness.

Mead was consumed in immense quantities around the world. It was enjoyed by kings, queens, saints, soldiers, and commoners. Mead was also the preferred drink of the Norse, Greek, and Roman gods, and it made mere mortals *feel* like gods. Queen Elizabeth I apparently loved her mead; her favorite recipe was a sweet mixture of spices, syrup, grape wine, and honey. Although mead flowed like rivers for centuries, it eventually lost popularity to other alcoholic beverages like wine, beer, port, and sherry.

Honey, wax, and mead. Honey bees must have seemed miraculous. Art was a way for people to worship and acknowledge the tiny insects that provided such great bounty. Across many cultures and across centuries, honey bees have been featured in paintings, carvings, jewelry, ceremonies, coins, tapestries, cloths, and flags. People have written about them in music, poems, and prose. Shakespeare gave a nod to honey bees in *Henry V* (Act 1, Scene 2), where the Archbishop of Canterbury describes how humans viewed bees:

> *So work the honey-bees;*
> *Creatures that, by a rule in nature, teach*
> *The act of order to a peopled kingdom.*

Honey bee behavior and biology has long fascinated us. Did ancient Spanish honey-hunters pause to observe and reflect upon the inner workings of the nests from which they stole? There are records to show that by the time of Aristotle, about two and a half thousand years ago, people were attempting to understand honey bee biology and behavior. In his *History of Animals*, Aristotle wrote, concerning where honey bees came from, "different hypotheses are in vogue. Some affirm that bees neither copulate nor give birth to young, but that they fetch their young." Specifically, people thought that honey bees grew from trees, especially olive trees. Despite a few other errors—such as bee colonies being ruled by a king—Aristotle wrote about honey bee comb construction, the division of labor in the nest, and swarming in impressive detail.

In 1609, Charls Butler's book *Feminine Monarchie* made a splash by insisting that the ruler of the hive was a queen, not a king. (That's not a typo in Butler's first name. Besides being an avid beekeeper, he was a strong proponent of English spelling reform.) The notion of queen bees caught on, perhaps because Butler's book was published just six years after the death of Elizabeth I, and people were warming up slightly to the idea of female leaders. In any event, Butler never explained how he came to his conclusion. However, a Spanish apiarist, Luis Mendez de Torres, wrote in his book *Tractado breve de la cultivacion y cura de las colmenas* twenty years earlier that the leader honey bee laid all the eggs which developed into workers, drones, and future queens. Some argue that Persian scientists knew about queen bees even earlier. Several decades after Butler's book, a Dutch anatomist named Jan Swammerdam dissected a queen bee and saw that she had fully developed ovaries with eggs and a sperm receptacle. Here was proof that what people thought were king bees were indeed queens. (Swammerdam's observations of insect anatomy also debunked the idea of spontaneous generation, such as the claim that bees grew from olive trees. He discovered that insects hatch from eggs laid by the female of the species. Another fun fact: Although Swammerdam was trained in human anatomy, he decided to use his skills to study insects instead.)

Through further dissections, Swammerdam confirmed that drones were males, and they impregnated the eggs that were inside the queen. But there was one problem: after examining their genitalia, Swammerdam couldn't see how the drones and queen would "fit." He concluded that the queen wasn't impregnated through copulation, but rather by "odoriferous effluvia"—a special smell or pheromone— that he called "*aura seminalis.*" A century later, François Huber, the blind Swiss naturalist, put this theory to the test. Huber and his assistant, François Burnens, removed all the drones from a beehive and ensured it only contained virgin queens. They then introduced a box of drones. The box had holes pierced in it so that the drones' *aura seminalis* could escape. However, this produced no offspring, suggesting that it took more than the drones' odor to impregnate a queen.

Then, one day in June 1788, Huber and Burnens saw a queen bee leave a hive and join a bunch of drones flying high up in the air. When she returned, Huber and Burnens discovered she was filled with seminal fluid. What they witnessed was the queen's mating flight: she leaves the hive to mate with drones, and then after mating, returns home. Huber and Burnens published their research in 1806, which resolved the debate about honey bee reproduction. (Interestingly, the Slovenian beekeeper Anton Janscha had witnessed the mating flight fifteen years earlier, but his account went mostly unnoticed.)

Honey bee colonies have queens and not kings? Honey bees are created from queens and drones having sex? These discoveries rocked popular thinking at the time. Yet honey bees had another surprise hidden up their sleeve: they were pollinators.

The discovery that insects were pollinators, and what this meant, was a long, drawn-out process. Ancient artwork shows Egyptian gods and Assyrian priests dusting date palm pollen from flower to flower, suggesting they knew something about fruiting plants and plant reproduction. Aristotle was onto something, too, when he saw honey bees carrying balls of "wax" on their back legs. However, the word "pollen," the Latin for powder or fine flour, wasn't used to describe the dust from flowers until the 1500s. One hundred years later, botanists realized that in order for plants to make seeds, this dust had to be transferred specifically from the male part of the flower to the female part. Jump to the late 1700s, and bees were identified as a way for this to happen. Then, in the late 1800s, Charles Darwin famously wrote about insect pollination systems and how important they are for the health and survival of a multitude of plant species.

The point of this extremely brief summary is to point out that much of what we know about pollination came *after* honey bees arrived in what is now known as North America. When those European colonizers brought *Apis mellifera* with them to the New World in the 1600s, it was for their honey and wax; however much we depend on honey bees for pollination today, that was not what we used them for then.

Colonizers kept honey bees in straw skep hives as well as wooden boxes or "log gums": sawed-off sections of trees. Honey bees that lived in trees came from swarms that had split off from colonies kept in skeps, boxes, or log gums and built their own nests in trees, as they had in the old world. Regardless of where the honey bees lived, harvesting honey was difficult and often deadly for the bees. People generally cut down the bee trees and took what honey was available, or gathered whatever honey was stored in a crude cap placed on the hive during the summer, or killed the bees and then removed the honey in the hive. To replenish their hives, they would capture swarms of honey bees in the spring.

In 1852, Lorenzo Lorraine Langstroth, a Congregationalist minister in Pennsylvania, patented a hive design that revolutionized beekeeping. His design consisted of a wooden box, taller than it was wide and deep, containing removable wooden frames, sort of like a wooden filing cabinet filled with hanging folders. The honey bees build the wax comb where they store their honey, eggs, and developing bees inside these frames.

Langstroth also discovered the importance of "bee space." This is a distance of about a quarter to five-sixteenths of an inch in width, just enough to let honey bees to crawl between the frames. Interestingly, this space is the same whether the comb is found in a man-made hive or in the wild. But Langstroth's design featured a departure from what the bees created for themselves. When honey bees build their wax comb in tree hollows, rock clefts, hollow logs, or skep hives, there are no inner supports. They construct and shape their comb to fit the open space. Sometimes the comb looks like suspended layers or walls, and sometimes the comb can end up being quite curvy and hilly, almost like a wide rollercoaster track. The vertical frames separated by bee space in Langstroth's design encouraged honey bees to construct their combs in a more "organized" fashion (at least by human standards). Often referred to as Langstroth hives, the good reverend's design remains the beekeeping standard today and was a major step forward in the evolution of commercial bee keeping. The idea of removable frames, however, came from François Huber. He designed frames so that they could be flipped like pages of a book, allowing the beekeeper to inspect the hive more easily. (Blind from a young age, Huber relied on his assistant, François Burnens, to carry out and document his experiments. This makes his insight into movable beehive frames all the more impressive.)

Langstroth hives also include what's called a "queen excluder," which is a metal grid that looks like a barbeque grate. Worker bees can squeeze through the bars but the queen can't, since she is bigger than her offspring. The queen is therefore confined to the lower layers of the hive where she lays her eggs. The upper layers are where the workers store the honey. This way, a beekeeper can harvest the honey without also harvesting eggs, larvae, and pupae.

Other inventions that followed the Langstroth hive helped make large-scale, commercial beekeeping and honey extraction possible. To calm the honey bees at honey-harvesting time, beekeepers created a "smoker": a hand-held tool that looks kind of like a small kettle. Beekeepers light a small fire inside it, and they can control how much smoke puffs out the nozzle. The invention of the centrifugal honey extractor allowed honey to be collected quickly and efficiently by spinning frames at high speeds instead of having to crush comb to squeeze it out. Bee veils evolved from pieces of cloth that beekeepers wrapped around their heads to protect them from stings.

For the next stage of our story we jump forward several decades from the development of the Langstroth hive. In 1891, a man by the name of Merton B. Waite was assigned by the United States Department of Agriculture to investigate why pear crops were failing in the eastern United States. The culprit was thought to be some kind of blight or disease spread by insects. To test this, Waite performed some experiments in pear orchards in New York and Virginia. These involved covering some of the buds of pear trees a day or two before the flowers opened, using mosquito net bags, paper bags, and bags made out of cheesecloth. The mosquito net bags were made of a very fine mesh that would keep out all insects but still allow wind and pollen to blow through. The paper bags kept out both insects and pollen, while the cheesecloth bags were considered somewhere in between the two. Waite kept the bags on the trees the entire time they were in bloom and after all of the flowers' petals had fallen off. Then he examined the quality of the resulting fruit.

What Waite discovered was that insect visits were not bad for the trees because they spread disease, but rather the opposite: they were needed for the trees to produce high-quality fruit. Bagged pear flowers produced poor fruit or none at all. Waite reported that he saw at least fifty species of insects visiting non-bagged pear flowers: bumble bees, honey bees, wasps, many species of sweat bees, ants, a large variety of beetles, fireflies, flies, an occasional butterfly, "and even a dragonfly." Waite saw that the pollen of pear flowers was too sticky and heavy to be carried by the wind, so it was up to insects to transfer it between pear blossoms.

Waite performed his experiments in 1891 and 1892, after much had been learned about pollination and after Darwin published his book *On the Effects of Cross and Self-Fertilization in the Vegetable Kingdom*. Drawing upon all this research, he realized that a number of common varieties of pear required insect pollination in order to produce quality fruit and seeds; specifically, these varieties needed cross-pollination between the flowers of different pear trees. For example, Waite observed that when Bartlett pear flowers received pollen from other Bartlett pear flowers (self-fertilization), the fruit was much smaller, had fewer seeds, and did not even *look* like a typical Bartlett pear. However, when Bartlett pear flowers received pollen from the flowers of Anjou, Easter, and other pear varieties (cross-fertilization), the fruit was heavier, had a plumper shape, and contained many more seeds. Insects, and especially bees, were carrying out this cross-pollination through their foraging trips to the different varieties of pear trees in the orchards.

Although Waite saw a wide variety of insects visiting pear flowers throughout his experiments, he wrote that, "The common honey bee is the most regular and important abundant visitor, and probably does more good than any other species." The honey bees that Waite saw were descendants of those brought to the United States roughly two hundred years earlier. And likely came from managed hives nearby or colonies of escaped honey bees nesting in trees or other cavities.

Significantly, Waite mentions that, "The sweat bees of the genus *Halictus* and *Andrena* are very abundant and useful." Based on research published in recent times, it would not be surprising if these native species of bees were the ones helping to produce the high-quality pears Waite saw, while the honey bees—the new bees on the block—were taking advantage of the flowers as a food source without contributing as much to cross-pollination.

Waite's observations are thought to have sparked the idea of moving managed honey bee hives to flowering crops to ensure pollination and good yields. In any event, beekeepers eventually added "pollinator" to their list of reasons to keep honey bees, and a new business opportunity was born. In her book *Bees in America: How the Honey Bee Shaped a Nation*, Tammy Horn describes how the pollination industry in the United States grew alongside the development of transportation. Initially, beekeepers sent their hives to farmers throughout the year using horse-drawn wagons. However, horses could be stung by the bees. Trains offered a way to move large numbers of hives quickly, but rail workers were often horrified to haul bees, and if railroad cars were left in the sun without ventilation, the bees were cooked to death. Nephi Miller, a beekeeper based in Utah, was known for leaving the railroad cars open and riding along with the bees to California, where they pollinated orange groves.

Fast forward to today, and cars, trucks, and interstate highways have made transporting beehives much easier. (As one editorial in the *American Bee Journal* stated, unlike horses, cars and trucks "will never get frightened, run away and break things by being attacked by cross bees.") Still, cross-country transport can be quite stressful for honey bees and is not without incident. For instance, beehives continue to fall off trucks on occasion, spilling millions of honey bees across roads and highways.

Despite the relationship humans have developed with honey bees over thousands of years, *A. mellifera* is, as mentioned before, only one of the approximately 20,000 species of bees that live on this planet. What has our relationship been like with the other 19,999? With many of these species, we have lived alongside them for centuries without any knowledge of their existence.

Chapter Two

L ET'S EXPLORE A bit more the landscape of ancient Spain where our naked friend was gathering honey. All around us is a visual symphony of wildflowers. Flitting about these, we can see *A. mellifera*—the bee from which our friend was stealing honey— with its rather slim body and slightly dark yellow and black color. It has delicate fuzz on its thorax—the middle segment of its body where its legs and wings are attached. If we pay attention, however, we will also see a variety of other species of bees in different colors, shapes, and sizes. Let's spend some time on them.

We will likely see solitary bees—species of bees that do not live in large colonies like *A. mellifera*, but on their own. Most of the world's bee species fall into this category.

As their name implies, solitary bees live a rather independent life. After a female solitary bee mates, she makes a nest, which includes individual "cells" called brood cells. She then stocks these with food (pollen and perhaps some nectar, depending on the species), and lays an egg about the size, shape and color of a grain of white rice in each cell. Then she leaves the nest and eventually dies, never seeing her offspring hatch and develop into adult bees. From the eggs she leaves behind hatch male or female bees; there are no queen solitary bees and no worker bees. They fly off to find food, and mate, and the cycle starts over again.

Where solitary bees build their nest depends on the species. Most make them underground, burrowing tunnels in bare, dry, light soil. Or they move into nests that were made and previously occupied by other animals such as ants, wasps, beetles, or other solitary bees. Some species build their nests out of mud, chew holes into wooden structures, or use pre-existing holes in trees, hollow stems, or other objects. A solitary bee nest can be found all on its own, isolated from other nests, or be one among a group of nests made by other solitary bees of the same species. But in these cases, their female owners do not appear to help each other. Each bee keeps to herself.

Solitary bees come in a mind-boggling variety. Some are so furry that they might be mistaken for bumble bees. Others are shiny like jewels and could be mistaken for flies. Solitary bees come in a huge range of colors, too, from bright green to purple, and a rich indigo blue.

Hawaii is home to solitary bees that are very different from the fuzzy, black and yellow bees that most people are familiar with. *Hylaeus* bees look like tiny black wasps. They are typically about six millimeters long—about the length of your little finger nail—and are virtually hairless. *Hylaeus* bees are also called yellow-faced bees because of the male's coloration. They are thought to be descendants of one single ancestor that arrived on the Hawaiian islands from Japan or East Asia roughly a million years ago. Over time, sixty different species of Hawaiian yellow-faced bees have evolved, an impressive diversification of an animal in such a, relatively speaking, short amount of time.

We will meet up with Hawaiian yellow-faced bees again a little later.

Returning to the ancient landscape of our honey-hunting friend, we would also see bumble bees: large bees covered in fur that might seem a bit clumsy, appearing to "bumble" from flower to flower, and whose buzz we can often hear above all the others. Their life cycle can vary depending on where they live. In temperate regions, after a queen bumble bee mates at the end of summer, she digs a tunnel underground or perhaps buries herself under material such as fallen leaves, and sleeps for the winter in a state of torpor. When spring comes, she digs her way out. Just like a bear emerging from hibernation, she is hungry and needs to fuel her body after such a long rest. She searches for nectar and pollen from flowers, which is why the availability of early flowers and blooming trees in spring is so important. She also searches for a place to establish her nest. If you see a big, fat bumble bee in spring, and she's flying low to the ground, chances are it's a queen looking for a home. They tend to choose existing cavities underground, such as abandoned rodent nests, but they may tuck their nest within any potential hiding place: clumps of long grass, human-made birdhouses, or even objects lying around in a yard.

Once the queen has found a nesting spot, she forms a ball of pollen mixed with her saliva and regurgitated nectar, lays eggs inside it, and covers it with wax. The pollen comes from flowers she visited; the wax is made by glands within her body, and she squeezes the wax out from her abdomen. The queen may also build a wax pot where she can store nectar for herself. Like a mother hen, she sits on her brood to keep them warm. She may sip from her nectar pot now and then. The queen also shivers her flight muscles to keep herself warm.

Like the eggs of solitary bees and honey bees, each egg that the bumble bee queen lays within the ball of pollen and wax is the size, shape, and color of a grain of white rice. Each egg hatches into a larva—a small, worm-like creature that is the second stage in bee development—and each larva spins a cocoon around itself. It is inside this cocoon that the larva grows into a bumble bee. In the spring and throughout most of the summer, the larvae develop into female worker bees. Near the end of summer, the queen lays eggs that develop into new queens and male bumble bees.

After the queen's daughters (worker bees) emerge, they take over the task of leaving the nest to find pollen and nectar from flowers. Pollen will feed the larvae; nectar, the queen and other adult bees. The queen remains in the nest, laying more and more eggs, which will result in more and more worker bees. When worker bees are not outside the nest finding pollen and nectar, they assist with "housekeeping" tasks. They can be sitting on the eggs and larvae to keep them warm and buzzing their wings to keep the nest temperature steady. They build more nectar pots to store nectar for when the weather is too poor to go out to forage.

Each worker bee lives for about several weeks, but the workforce is continually replaced by new bumble bees that emerge throughout spring and summer. There are "undertaker bees" that carry dead bumble bees from the nest and dump them outside. This not only keeps the nest clean, but it also provides food for animals that eat bees. If the dead bee is not eaten, its body breaks down and provides nourishment for the soil.

Around late summer, the eggs that the queen bumble bee lays hatch into new queens and male bumble bees. Once the new queens and males are fully grown, they leave the nest to mate. Male bumble bees die after this. The original queen and all the female worker bees eventually die, too. After a new queen mates, she buries herself in the ground where she will stay for the winter, just like her mother did. Then the cycle begins again when she digs her way out in the spring.

There is diversity in bumble bee lifecycles, however. Some bumble bee species, such as the early bumble bee (*Bombus pratorum*) in the United Kingdom, produce two generations in spring and summer. In parts of the world where there are warmer winters, such as in Mediterranean regions and New Zealand, bumble bees can be active and produce generations in autumn and winter. Surprisingly, in the past few decades in southern England, and increasingly further north in that country, naturalists have seen buff-tailed bumble bees (*Bombus terrestris*) throughout the winter months when they would usually be hibernating. Thanks to milder winters and urban parks and gardens that provide food-rich, winter-flowering plant species like Mahonia, winter honeysuckle, winter heather, and snowdrop flowers, some buff-tailed bumble bees are able to establish a second generation when most other bumble bees are overwintering.

Of course, bumble bee species that live in tropical areas don't have to contend with cold winters and some species are active all year long, and some colonies persist for more than one year. The lifecycle of the Amazonian bumble bee, *Bombus transversalis*, appears to be synchronized with the cyclical wet and dry seasons: queens establish their colonies during the wet season and die off in the dry season. Instead of using pre-existing underground nests like bumble bees elsewhere, these bees build their nests above ground in areas unlikely to flood in the rainy season and weave a sort of thatched roof over them.

At least 250 species of bumble bees have been identified. Interestingly, however, there are more than 400 different bumble bee color patterns, various combinations of black, white, yellows, oranges, and reds combined on the bees' bodies in myriad ways. The darkest bumble bees are usually found in the tropics, the palest, in the intermediate northern latitudes, such as in the United States, Canada, and the United Kingdom. Their colors are thought to blend in with grasslands. Some species are strongly banded or striped, while others might have a more solid color. Bumble bees come in a variety of fur patterns that include not only the familiar yellow and black, but also bright orange, snowy white, and rusty red. Each species has its own distinctive fur-color pattern, although fur colors and patterns can also vary within some species. For instance, in some bumble bees, males have different-colored fur than the females, and the queens of some species have a fur-color pattern that differs from their offspring. Sometimes the differences in bumble bee fur patterns are so subtle that it's difficult for experts to identify a species in the field before it flies away. The one sure way to identify a species is DNA testing.

To return our attention to ancient Spain once again, the people who lived there might have noticed a variety of different bees. However, *A. mellifera* made an unforgettable impression with the sheer amount of honey it can store. Even if people discovered nests of other bee species, the amount of stored nectar they would find would pale in comparison to what *A. mellifera* could provide. With colonies composed of tens of thousands of worker bees, *A. mellifera* must store copious amounts of honey to feed all those mouths and keep them going over the winter. Other bee species in ancient Spain would have lived in much smaller colonies, and thus have needed to store much less nectar (Nectar is the sugary fluid that flowers secrete to attract pollinators. Honey bees turn nectar into honey—which is sweeter and more viscous—by adding their own enzyme and evaporating the water content.) No wonder that *A. mellifera* was the subject of choice for the paintings in La Araña, and that later the Western world would shine a spotlight on *A. mellifera* over all other species of bees.

However, honey-hunting in Paleolithic times was not limited to Spain, or Europe, or to *A. mellifera*. Rock paintings in Africa, India, and Australia show that ancient peoples around the world pillaged bee nests for honey. The bees would have been different species depending on the location. In Asia, honey was available from *Apis cerana*, the Asian or eastern honey bee. In Africa, honey was provided by *Apis mellifera scutellata*, the African honey bee, or *Apis mellifera capensis*, the Cape honey bee. (*Apis* is the name give to the genus of bees considered to be true honey bees. A species' specific name comes after *Apis*.)

People have also gathered honey for thousands of years from stingless bees. These bees, also called meliponine bees after their scientific name *Meliponini*, are found in tropical or subtropical areas, including South America, Southeast Asia, Africa, and Australia (where they are also known as "sugarbag" bees). There are at least 300 different species of stingless bees, in a range of sizes, shapes, and colors. They can be as tiny as a pea or as big as a grape. They boast a range of colors including gold, onyx, garnet red, and cinnamon, with stripes or without. Some species have eyes that are beady black, whereas others have eyes that are bluish-gray or peridot green.

Stingless bees live in social groups like honey bees and bumble bees. Depending on the species, colonies can have tens of thousands of worker bees, which usually means they store lots and lots of honey. As their name implies, they do not sting. Imagine being able to remove honey from a bees' nest without being stung! It is no wonder that to this day, stingless bees are known as being good-natured animals and are referred to as "little angels" (*angelitas*) or "little maidens" (*doncellitas*). However, they are not entirely defenseless: if needed, they can use their mouthparts to deliver a painful bite.

Across Europe, especially in Nordic countries, bumble bees played a starring role. Children searched for bumble bee nests so they could suck out the nectar using a drinking straw made of grass. Historical records note this happening in the 1800s, when sugar was still a luxury, but bumble bee hunting in Europe was likely an old tradition stretching far back in time. As mentioned, bumble bee nests tend to be well hidden in tree trunks, tussocks, or abandoned rodent nests underground. It is possible that ancient peoples spotted bumble bees flying in and out of tangled heaps of grass, or from a tree trunk. Curious, they parted the blades or looked inside the hollow to see what was there. What they would have seen was clusters of jellybean-sized cocoons, wax-covered eggs, and tiny wax pots filled with nectar, all arranged in a jumble. Depending on the species of bumble bee, they would have seen anywhere from 50 to 300 worker bumble bees sitting on the brood, keeping the eggs and cocoon-encased larvae warm. With fewer numbers and no need to store food for the winter (most species of bumble bees die off in the fall and the surviving queen hibernates underground until spring), bumble bees do not store much nectar—only enough for a rainy day or two. They also don't convert stored nectar into honey, so the liquid people found in a bumble bee nest would be more watery and less sweet than honey. Nevertheless, it would have been a tasty treat.

Bumble bee nests were often found during haymaking because they can hide their nests nicely inside piles or bales of hay. European herdsmen walking in the shrub landscape looked for bumble bee nests so they could bring home what they called "bumble bee honey" or "bumble bee mead." Their families would eat the nectar with their morning porridge or as a dessert after supper. A visitor to that time might see a young lad offering bumble bee nectar to a girl as a romantic gesture, or a mother putting bumble bee nectar into the ear of her child in an attempt to soothe a pain.

People relied on traditional knowledge to find nests and the different types of bumble bees. If we were to explore the Nordic landscapes of the 1800s and earlier, we would see a plethora of species. Even today, Europe in general is home to a range of bumble bees. For example, Denmark has about twenty-nine species, Finland, thirty-eight, Estonia, nineteen, Norway, thirty-six, Sweden, forty, and Iceland, four. All of these different species have different appearances and nesting characteristics. The heath bumble bee (*Bombus jonellus*) has fluffy yellow-black-yellow bands down its body, with a tuft of white fur on its rear end. Their nests are rather small, containing about fifty workers. On the other hand, the red-tailed bumble bee (*Bombus lapidarius*), as its name suggests, has a bright red or orange end that contrasts with its overall black fur. They create fairly big nests containing several hundred workers. People long ago could likely recognize the bumble bees with relatively large nests, and thus a larger bounty of sweet nectar.

In Shakespeare's time, bumble bees were often referred to as "humble-bees." In *A Midsummer Night's Dream* (Act 4, Scene 1), Bottom says,

> *Monsieur Cobweb, good monsieur, get you*
> *your weapons in your hand and kill me a red-hipped*
> *humble-bee on the top of a thistle, and, good*
> *monsieur, have a care the honey-bag break*
> *not; I would be loathe to have you overflown with a*
> *honey-bag, signior.*

Bottom is likely referring to the red-tailed bumble bee, which would have been spotted in meadows across Europe.

Charles Darwin was introduced to "humble bees" by his young son, Georgy. When Georgy was eight or nine, he noticed bumble bees flying the same path over and over. When Darwin and Georgy followed the bumble bees, they saw that each bee would stop and hover for a few seconds at particular points along their route: first at a bare spot at the side of a ditch, and then at a spot several yards down over a particular ivy leaf. Darwin called these spots "buzzing places." He also noted that all of the bumble bees pausing were male *Bombus hortorum*, the garden bumble bee. (Fun fact: the garden bumble bee has a particularly long tongue or *proboscis* and can sip nectar from flowers with particularly deep nectar tubes.) Besides garden bumble bees, Darwin and his children (he had many and they were all drafted into searching for bumble bees) also observed male *Bombus pratorum* (early bumble bees) and *Bombus lucorum* (white-tailed bumble bees) behaving the same way.

Darwin and his children studied buzzing places for several years, from 1854 to 1861. Why they abandoned their project remains a mystery, and Darwin never published his observations. The details come from Darwin's field notes and Georgy's memories recalled many years later when he was Sir George, a distinguished astronomer. The buzzing places remained a puzzle. At one point in his field notes Darwin wrote, "Is it like dogs at corner stones?" Just as dogs pee to mark their territory, leaving a scent message for other dogs, do male bumble bees mark buzzing places with a scent for other bees?

Many years later, scientists discovered that Darwin wasn't far off. Buzzing places are in fact marked with a pheromone that comes from the male bumble bees' *labial gland*, found in its head. The male bumble bees "paint" the pheromone onto objects, such as leaves or sticks or trees, using the bushy mustache around their mandibles. This mustache makes male bumble bees look like they have a big, fuzzy nose, which is an easy (and charming) way to identify them.

These pheromones differ among different species of male bumble bees. Also, bumble bee species differ in terms of the height at which they create their flight paths and buzzing places. For example, male garden bumble bees tend to fly close to the ground, which would have made it easier for Darwin and his children to spot them.

What could be the point of buzzing places? Darwin and scientists since his time believe that male bumble bees use them to attract a mate. The pheromones that male bumble bees paint on the buzzing places could attract queen bumble bees that need to mate before they hibernate for the winter. By patrolling the same route over and over, male bumble bees might eventually run into a queen. The different pheromones and different flight heights between species are thought to exist so that male bumble bees mate with their same species of queen.

However, the funny thing is that no one, not even Darwin and his children, has ever reported seeing a queen bumble bee hanging out at a buzzing place.

Now let's turn to North America. As mentioned before, *A. mellifera* did not exist there before European colonizers arrived. However, there was an abundance of native solitary and bumble bees. Like all plants and animals, these bees were regarded by native peoples as integral parts of the web of life. Traditional knowledge about native bee species was passed down from generation to generation. Dr. Henry Lickers, an Elder of the Seneca Nation, Turtle Clan, Haudenosaunee (also known as the Iroquois), shared with me the following story and traditional knowledge about the bumble bees with whom they shared their land. The Haudenosaunee were the first agricultural peoples to live in what is now northeast Canada and the United States. They continue their practices today.

"I grew up on the Six Nation Reserve near Brantford, Ontario, Canada. I remember there was an old barn that they used to store hay bales in, for bedding for the cattle. One day I was in the barn and I saw a bumble bee fly out of one of the bales of hay. Then another, then another. Then I saw one fly back in. I watched for a while as they flew in and out.

"I didn't know anything about bumble bees, but I was curious, so I grabbed a stick and approached the bale of hay. I used the stick to pry it apart little by little. Sure enough, didn't I come to a great group of [what looked like] . . . I want to call them gray, gray to white grapes, in a little bunch. The bees got a little upset, so I put the hay back and went to see my grandmother, who knew about these things.

"I told my grandmother what I saw, and she asked, 'Did you sing the bee song?' And I said, 'Grandma, I don't know the bee song.'

"'I will teach you the bee song,' she said.

"So, my grandmother followed me out to the barn. I noticed that as I approached the spot where I had seen the bumble bees, my grandmother stayed back. 'You go forward,' she instructed, 'And me as an adult will stay back. That is the proper way.'

"My grandmother told me to sit near the bees and wait and listen. I had to listen very carefully to how the wings of the bees were flapping. Sure enough, after a while I could hear it. It was like a humming sound. Some wings were flapping really, really fast, and others were flapping really, really low. Together, a type of harmony

was created, or a chorus of notes. And as the bees worked, some were flapping their wings while others weren't. As I listened really carefully, I could hear what the song was. 'Now hum along,' said my grandmother. And as I hummed along, the bumble bees became very calm. 'Every bumble bee nest has its own song,' explained my grandmother. 'You have to listen for it.'

"As I hummed along with the bees, it was as if to them I wasn't even there. Or, it was as if I was another bee. I started to slowly take the nest apart, being careful not to just rip it apart. You have to take a really slow time to look. 'You'll see these things that are about the size of a big pea,' my grandmother said. 'When you see those, don't reach in and grab them all. You grab just one. The bees might start flying around a little bit more, but you keep the humming going. The bees are a peaceful tribe. The bee makes that honey for her own kids and makes it for <u>any</u> kids. But unless you treat it with respect and hum to them, they won't give it up easily.'

"'You put that little pill in your mouth,' my grandmother continued, 'hold it, and put the nest back together again. Then you can back away, squash the pill in your mouth, and eat the whole thing.'

"As I backed away from the nest, I squashed the 'pill' in my mouth with my tongue. It felt like . . . I want to say it's like a gelatin coat. You know how Jell-O goes when it gets hard? It's like that. And of course, when I squashed it, the honey came out. But my grandmother never called it honey, though. She called it something else, like a honeydew, or a dew type of thing. She said that it's sweet, like maple sap. And as I tasted it, I realized she was right.

"After that day I found a number of bumble bee nests. They really interested me. There would have been about three or four acres of

bush and trees, and I usually found them around the edge, where you had thick grass. I found them in there. And there was usually some type of structure with it, like a rock. So, they would be beside the rock but in the grass, and you could see how they were sort of pulling the grass down to it. I remember one nest was at the edge of a marsh, in an old muskrat house. But I never told anyone where I found the bee nests. I kept it a secret. And I always left fresh grass very near to the nest for the bumble bees."

Over the years, some people kept bumble bees. After finding a bumble bee nest hidden in a meadow, people in Transylvania would put the nest into a cylindrical earthenware "bumble-house" that had a cover. They would leave the bumble-house until the evening so that any foraging bumble bees had time to return to their new home. The bumble-house was then brought back to the village, where it was placed in the shade and people would harvest the nectar now and then. Village houses often had three to four bumble-houses.

An Englishman by the name of Frederick William Lambert Sladen longed for "complete domestication" of bumble bees. By this he meant that they would become pets and live with humans and depend upon us, like cats and dogs. In 1912, he published *The Humble-Bee: Its Life History and How to Domesticate It*. In this book, Sladen provides impressive details about the life cycle of a bumble bee colony, how to distinguish the various species of bumble bees found in Britain, and recounts his many entertaining attempts to observe and raise bumble bees on his own. He constructed what he called a "humble-bee house": a six-foot by four-foot wooden shack in his back garden. Inside the humble-bee house were two shelves, one on each side, extending along the building's length. On each shelf there were four wooden boxes with a glass top that allowed him to see inside. In each box, Sladen placed a bumble bee nest that he had dug up from his property. There was a tube from each bee-box leading to the outside, so that the bumble bees could come and go freely to collect food. A window in Sladen's humble-bee house gave him enough light to watch the bees during the day. When he was not watching them, he covered the window with a thick, black blind. This was to mimic the darkness of natural underground bumble bee nests. In the evenings, Sladen observed the bumble bees by candlelight.

At one point, Sladen went so far as to keep one of the wooden boxes of bumble bees on a table in his study. If the weather outside was nice, he left a window in his study open during the day and the bumble bee box open, so that worker bumble bees could come and go as they pleased. The bumble bees would fly straight to the window when leaving and fly straight back to the box upon their return. When the worker bees returned, he noted he could often see large balls of pollen on their back legs. Sladen closed the study window in the evening once all of the worker bees were back at the nest. If it was very windy outside or raining, Sladen would keep the nest and study window closed for the day and give the bumble bees some honey to tide them over until the weather improved and the bees could go outside to forage for nectar for themselves.

Sladen was very successful in learning how to observe and care for bumble bee colonies outdoors, in his humble-bee house, and inside in his study—and he provided many details in his book that would allow others to follow in his footsteps. Still, the piece of the puzzle that eluded him was how to mimic or prevent the hibernation that queen bumble bees undergo over the winter so that he could study them all year.

Fast forward to the early 1990s. In the Netherlands is Koppert Biological Systems, a natural crop treatment company, and in Belgium, Dr. Roland de Jonghe, a veterinarian with a keen interest in bumble bees. Both the Kopperts and de Jonghe solved the mystery of how to avoid the need for bumble bee queens to hibernate, thereby allowing them to breed bumble bee queens year-round. How they do this is still a tightly kept secret. de Jonghe started his own company, Biobest. Both Biobest and Koppert Biological Systems set up factory-like conditions to mass-produce bumble bee colonies. Each bumble bee colony is placed in a plastic box with a cardboard outer box, given a supply of pollen and sugar water (as a substitute for flower nectar), and then sealed up and mailed to its destination, usually tomato greenhouses. Bumble bees are used to pollinate a number of other fruits and vegetables, too, in greenhouses and in open fields.

There is a high demand for tomatoes (think of all the pasta sauce, pizza sauce, ketchup, and salsa people consume), and bumble bees are superstar tomato plant pollinators. Tomatoes hide their pollen inside their anthers, and to get it out, the anthers have to be vibrated or shaken. Bumble bees do this by grasping the flower with their legs and mouthparts, curling their body around the flower, and vibrating their flight muscles really fast. This creates a high-pitched buzz sound, which led to it being called *buzz pollination* (also known as *sonication*). Vibrating their flight muscles causes the pollen to fall out of the anthers, sort of like shaking salt out of a salt shaker. The pollen falls onto the bumble bee's belly, and she uses her front and middle legs to scrape it and gather it onto her back legs. Any pollen grains that she missed can rub off on the next tomato flower she visits. Usually the bumble bee leaves little brown marks on the anther after it has been buzz pollinated and commercial tomato growers call these "bee kisses."

Tomatoes are not the only plant that can be buzz pollinated. Eggplant, kiwi, blueberry, cranberry, and thousands of other plant species have poricidal anthers like tomato flowers. Honey bees do not buzz pollinate, making bumble bees and other types of bees that can the preferred pollinators for these types of plants. Tomato plants that are buzz pollinated by bumble bees tend to produce higher quality tomatoes compared to those that are not. In the absence of bumble bees, people can hand-pollinate tomato flowers, moving from plant to plant, vibrating each flower using a handheld device that looks like a big electric toothbrush. This is very time-consuming and expensive. Bumble bees are a much more natural, simpler, cost-effective strategy.

Once the mystery of how to rear bumble bee colonies was solved, the commercial bumble bee industry was born. Colonies could be shipped where they were needed. Producers in Europe breed and export one of its native species, *Bombus terrestris*, the buff-tailed bumble bee. Canada and the United States focus on the native *Bombus impatiens*, the common eastern bumble bee. By 2004, sales across the industry reached one million colonies. Sales figures after 2004 is another tightly kept industry secret. No recent sales reports from the industry exist. However, nothing indicates that sales have slowed. As we will see, this boom in commercial bumble bees over the years has come at a cost.

Chapter Three

L ET'S RETURN OUR focus to honey bees. Our naked Spanish friend from ancient times would be blown away by the vast quantity of honey we harvest these days. The ancient Egyptians, Greeks, Romans, and people in Medieval times would be astounded at the scale on which honey bees are currently managed, and how much people rely on honey bees for both food and income. Once beehives gracefully floated down the Nile; today large trucks transport multitudes of hives cross country. Beekeeping has also become quite varied. At one end of the spectrum are people who manage only one or a few colonies for pleasure. Then there are others who manage a larger number of hives, anywhere from tens to hundreds of hives, with their colonies being an important source of income from selling honey and other honey bee products, and from renting out bees to pollinate crops. At the far end of the spectrum is what some call production agriculture or commercial beekeeping: huge operations that manage thousands of colonies, trucking their honey bees hundreds of miles each year to pollinate the food crops that humans depend on. Urban beekeeping companies, installing and managing honey bee hives in major cities, have also entered the scene.

To give a sense of the scale of modern beekeeping, the Canadian Honey Council estimates that in Canada alone, there are approximately 13,000 beekeepers and a total of about 810,000 colonies of honey bees. Most of these colonies are managed by commercial beekeepers rather than those keeping bees for a hobby: about 20 percent of the beekeepers maintain 80 percent of the colonies. The average number of honey bee colonies that commercial beekeepers maintain depends on the region. Commercial beekeepers in the prairies maintain anywhere from 500 to 13,000 colonies each, with an average of about 2,000. In eastern Canada and British Columbia, commercial beekeepers manage from 50 to 5,000 colonies, with 600 colonies being the average. The Canadian Honey Council estimates that the value of honey bees to the pollination of crops is more than $2 billion annually. Then there's honey. Canada produces about 75 million pounds each year, and exports about half of that, mostly to the United States. According to Agriculture and Agri-Food Canada, in 2022, the value of honey produced in Canada was about $254 million.

In 2022, the United States was estimated to have 2.88 million honey bee colonies. Between July 1, 2017, and January 1, 2018, more than half a million managed honey bee colonies were transported across the United States to pollinate the almond fields in California, a crop that is worth more than US$5 billion. Honey bees are big business.

I spoke with a former employee of a large commercial beekeeping operation that provided pollination for blueberries. "He had 10,000 hives, I think," they told me. "I managed 5,000." I thought I didn't hear them correctly. "You managed five *thousand* hives?" I asked. Yes, they told me, five thousand. They described what it was like preparing so many beehives for travel. "I think it was like 700 and something hives per truckload, just getting pitch dark because you have to load the bees in the dark. So, the forklifts have red light because the bees can't see . . . are not supposed to see red. And just moving pallets of bees and stacking them onto this truck and you just look around and you're like, 'Oh my, this is insane.'"

"We would offload them off the truck in the middle of the night," they continued, "and then the next morning another crew would come break the hives in half, like bottom stays here, top goes here, put them out to bee yards the next night, but as we were breaking them apart of course they're [the bees] all waking up and they're like, 'Where the hell am I? I'm supposed to be in New Brunswick and now I'm in Ontario,' and like, 'The top half of my hive is gone' . . . So the bees are just swirling around and you just, you got used to it that you're just walking through a cloud of bees."

The scale of modern-day beekeeping is by any standard stupefying. We have certainly benefited. But achieving this has come at a price. We're paying for it, and worse, so are the bees. "The bee declines these days are a big concern," Lickers told me. "You know, it's like losing one of your relatives. To our people, it shows we haven't respected them. . . . And so, the bees are leaving."

Hackenberg Apiaries, based in Pennsylvania, is an example of a large commercial beekeeping operation. In addition to selling honey, each year Hackenberg Apiaries sends its beehives across the

United States to pollinate various crops. And the bees are on a tight schedule. From February to mid-March, the honey bees are needed in California to pollinate almond fields. Then, from mid-March to mid-April, the honey bees are in Florida taking care of citrus fruits. Next are apple crops in Pennsylvania and New York. By mid-May, Hackenberg's honey bees are in Maine visiting blueberries. There is a bit of a break in June when the honey bees hang out by the St. Lawrence River, where they can focus on making honey. In July, the road trip resumes: Pennsylvania pumpkin patches until mid-August, then Florida to make honey from Brazilian peppers. By February, the honey bees are back in California with the almonds, and the itinerary repeats. Year after year. The annual round trip, all by truck, is about 17,700 kilometers.

The financial stakes are high. David Hackenberg, founder and owner of Hackenberg Apiaries, estimates that the almond crop alone is valued at US$900 million, citrus fruits at about $800 million, apples at more than $1 billion, blueberries at about $150 million, and pumpkins at around $200 million. The stakes are also high for human nutrition. These crops contribute to a balanced diet of essential vitamins and nutrients and provide many of the foods we enjoy. Each of Hackenberg's 2,000 hives contains roughly 20,000 worker bees (that's a conservative estimate). He helps the United States meet its pollination and honey production needs by providing a workforce of approximately 40 million honey bees.

Besides being known for its honey and migratory beekeeping, Hackenberg Apiaries was also the "epicenter" of what has so far been the biggest and most mysterious phenomenon the beekeeping world had ever experienced. It was fall of 2006. David Hackenberg, his family, and his honey bees were in Florida, as per their schedule for the past forty-two years. When Hackenberg did his routine check on his 3,000 hives, he was pleased to see they were "boiling over with bees." A month later, during another routine check, he was smoking his hives when he noticed there were no honey bees flying around. Instead of being greeted by the constant, familiar, soothing hum of thousands of honey bees, he was surrounded by . . . silence. He opened up his hives. To his horror, his honey bees were gone. Except for the queens, some drones, and a few young worker bees, more than half of his 3,000 hives were empty. There were no signs of conflict. There were no dead or dying honey bees anywhere. Tens of thousands of honey bees had simply vanished, leaving their copious stores of honey behind. "It was like a ghost town," Hackenberg said.

Seeking answers, Hackenberg picked up the phone. One of the experts he called was Jerry Hayes, at the time the Chief of the Apiary Section of the Florida Department of Agriculture. "Beekeepers are always connected," Hayes explained. "We're a small family, and I was hearing a couple of beekeepers, I think in South Carolina or Georgia, saying that they had lost all their colonies. The bees had disappeared." Like Hackenberg, these beekeepers found no dead bees. "The bees were just gone," Hayes said, "and I thought to myself, 'Yeah, whatever,' because beekeepers were having all sorts of problems back then." When Hackenberg, who had a particularly large beekeeping operation, told him his bees were gone, Hayes thought, *No, that can't be. That can't be . . .*

Hackenberg convinced Hayes to come out and look at his hives. "He was absolutely right," Hayes said. "The bees were gone. Queen was still there but the bees were gone."

And the bees didn't come back.

By the following spring in 2007, only 800 of Hackenberg's original 3,000 hives had survived. (These days they are back up to between 1,500 and 2,000). More than 40 million of his honey bees had vanished. Hackenberg reached out to other beekeepers across the United States and learned that he was not alone. The media picked up on the mystery. The *New York Times* announced, "Bees Vanish, and Scientists Race for Reasons," and "Honeybees Vanish, Leaving Keepers in Peril." Reuters proclaimed, "Vanishing Honeybees Mystify Scientists." "Where Have all the Honey Bees Gone?" asked the *New Scientist*. The disappearance of honey bees was described as "a mystery worthy of Agatha Christie." The underlying concern was captured in a headline by ABC News: "Honeybees Dying: Scientists Wonder Why, and Worry About Food Supply." If honey bees were disappearing, what would happen to all the crops they pollinate? What would happen to all the almonds, blueberries, fruits, pumpkins, and peppers that relied on Hackenberg's bees and bees from other beekeepers? A statistic began circulating that one third of the food we eat relied on pollination. Parts of Europe were reporting honey bee disappearances, too. Amidst the simmering panic, a number of potential culprits emerged: genetically modified crops, cellphone towers, high-voltage transmission lines, a secret plot by Osama bin Laden to bring down American agriculture, and the "rapture of the bees," in which God was recalling them to heaven.

The media were fixated on honey bees, but they were not the only bees that were disappearing.

Robbin Thorp knew the trails well in Mount Ashland, Oregon. A retired entomologist from the University of California-Davis, he walked them regularly, with a butterfly net in one hand and a "bug vacuum" in the other. The bug vacuum looked like a big water gun, and when Thorp pulled its trigger, it sucked up whatever insect he was aiming it at. The insect ended up in a transparent section of the bug vacuum where he could examine it more closely. After he identified the species, he set it free.

But Thorp wasn't looking for just any insect. He was searching for Franklin's bumble bee (*Bombus franklini*). To the untrained eye, Franklin's bumble bee might look like any other big, fuzzy, black and yellow bumble bee. But Franklin's bumble bee is distinguished for its round black face ("some of the others have a very long face," Thorp said). Its black abdomen (the large hind section of the body) stretches up into a U-shape around its yellow thorax (the middle section of the body), extending just beyond the base of its wings. Franklin's bumble bee looks a lot like the yellow-faced bumble bee (*Bombus vosnesenskii*), which, true to its name, has a yellow face (Franklin's has yellow on top of its head) and a distinguishing yellow stripe on its abdomen. The yellow-faced bumble bee could often be seen buzzing around the Mount Ashland trails; Franklin's bumble bee, not so much. Back in the 1990s, Thorp saw many Franklin's bumble bees every day. "I could walk down and see [them] on every patch of flowers." In 1998, he saw ninety-four Franklin's bumble bees. In 2002, he saw twenty. He only saw three in 2003. In 2004 and 2005, he saw none. "It was just gone," he said.

It was August 9, 2006, and Thorp was walking the trails of Mount Ashland once again. At seventy-two years of age, he could identify bumble bees in a split second, and on that day, he saw one that he instantly knew was a Franklin's bumble bee. By then, he hadn't seen any in three years. It zipped through a meadow and Thorp frantically raced after it. Alas, he never caught it; it disappeared across a field of yellow buckwheat. Thorp continued to search for Franklin's bumble bee up until his death in 2019. In 2006, the same year that Hackenberg discovered that millions of his honey bees had disappeared, Thorp saw his last Franklin's bumble bee.

The year 2006 also happened to be when scientist Sheila Colla wrapped up a large field study across southern Ontario, in Canada. Accompanied by field assistants, each wielding a butterfly net, they had caught as many bumble bees as they could find. When they caught one, they identified the species, marked the bee with a non-toxic bright-colored powder so that they would know it had already been counted, and then set it free. Colla and her colleague, Laurence Packer, compared all the data they collected to a large report on bumble bee populations published in the 1970s. To their dismay, they saw only one rusty-patched bumble bee (*Bombus affinis*), and there were two previously reported bumble bee species that they had not seen at all: the American bumble bee (*Bombus pensylvanicus*) and the yellow-banded bumble bee (*Bombus terricola*).

Colla was especially concerned about the rusty-patched bumble bee—named for the rust-colored patch on the yellow section of its abdomen. Thirty-five years earlier it was seen quite often, but now it had practically vanished. The species is closely related to Franklin's bumble bee, which she had learned from Thorp's work, had seemingly disappeared. Did the rusty-patched bumble bee live anywhere else besides where she had surveyed? If she looked in other areas, could she find it?

One way to determine where a bumble bee species can possibly be found is to use museum specimens or collections of bumble bees. Captured in the wild, the specimen is killed then skewered with a pin, which is then stuck through tiny pieces of paper on which are written the location where the bee was found, and the date, what type of plant the bee was feeding on, the species of bee, the name of the person who identified it and the date when they did.

Colla found a number of rusty-patched bumble bee specimens in various museums and collections in Canada and the United States, some dating as far back as 1903. After noting where the specimens had been found, she had a pretty good idea of the bee's historical range. It was time to hit the road.

Colla visited forty-three sites in total: eighteen in eastern Canada and twenty-five in the eastern United States. After surveying these, and catching and identifying about nine thousand bumble bees in total, she had found only one rusty-patched bumble bee. It was in Pinery Provincial Park in Ontario, Canada. "I was actually sitting in the passenger side of a car, looking out the window, and I saw it on the road," Colla told me. "I told my friend to stop because I saw a bee that looked a bit different. And then we caught it. That was in 2009. And that's the last one that's been seen in Canada. No one has seen it since then."

Before honey bees began mysteriously disappearing from apiaries across the United States, one question that was more specifically on the minds of a lot of scientists was: What was happening to the *world's* bees? Franklin's and the rusty-patched were just two out of a number of bumble bee species that had been declining precipitously across North America, South America, and Europe.

Besides Franklin's bumble bee and the rusty-patched bumble bee, a report published in the early 2000s suggested that the native yellow-faced bees on the Hawaiian Islands were in trouble, too. Scientist Karl Magnacca realized that not much was known about Hawaii's yellow-faced bees, so he did an extensive multi-year field study across all the Hawaiian islands. He found that of the sixty different species of yellow-faced bees on the islands, ten were possibly extinct. What was happening to Hawaii's bees? This didn't seem to be an isolated problem. Managed colonies of *Melipona beecheii*, a honey-producing stingless bee, showed a 93 percent decline in the Yucatán peninsula over the previous twenty-five years. Were the thousands of other bee species around the world suffering the same fate? There was not enough data available to make any firm conclusions. The headlines about vanishing honey bees revealed only a sliver of a much more alarming and complex state of affairs.

Chapter Four

T O UNDERSTAND WHY managed honey bees and some species of wild bees were disappearing, it's important to go into a bit of detail about the diseases faced by both managed honey bees and wild bees, how they spread, and the damage they do.

Some bee diseases can be seen with the naked eye—and smelled with the naked nose—whereas others are more difficult to spot, and might not be detected until it's too late. As in the case of livestock, such as cows and pigs, the way managed honey bees and bumble bees are kept crammed together in large numbers is unnatural and can create ideal conditions for disease to spread.

When David Hackenberg phoned Jerry Hayes in 2006 about his vanishing honey bees, Hayes already knew that honey bees were having "all sorts of problems." In fact, the history of commercial beekeeping in North America over the past more than one hundred years has been a story of one problem, one devastating problem, after another.

One major problem that has been around for quite some time is American foulbrood, a devastating disease that infects honey bee larvae.

American foulbrood is a spore-forming bacterium that is inadvertently fed to larvae by adult honey bees in the colony. After a larva is fed, honey bees seal the larva's cell with wax, creating a cozy little compartment where the larva can develop into an adult bee. In a larva infected with American foulbrood spores, the spores germinate in the larva's gut and multiply rapidly. The cap on the larva's cell becomes moist and darkens. As the larva shrinks and dies, the capping sinks down into the cell, which gives it a concave rather than convex shape. Worker bees, being diligent housekeepers, will notice the sunken capping and chew holes in it. Sometimes they chew the cell cap completely off and remove the dead larva. However, cleaning out the cell spreads spores (and the disease) from the dead body throughout the hive. More larvae become infected. Spores can also get into the stored honey in the hive, infecting any honey bee that eats it. As the disease spreads to more bees and the hive weakens, it cannot defend itself from robber bees from nearby strong colonies. Robber bees can take the contaminated honey back to their own hive, and that hive now becomes infected. Beekeepers can also spread American foulbrood by exposing honey bees to contaminated honey or hive equipment.

If you open a honey bee hive that is infected with American foulbrood, there is a nasty, sour smell—hence the name "foulbrood." The comb looks patchy with a mixture of dark, diseased cells and lighter-colored, healthy brood cells. There may be cells whose capping has been punctured or chewed away by worker bees, exposing a dark, dead larva. If infected larvae have been dead for a while, the cells will contain dark, scale-like remains. The beekeeper will see sunken, moist cell caps, and if they insert a stick into these cells, the larvae's remains can be drawn out.

If a colony is infected with American foulbrood, there is no treatment. It's doomed. To prevent the spores from spreading to other colonies, the infected colony should be burned—equipment, bees, and all. Over the years, many beekeepers have lost numerous colonies from this disease.

European foulbrood is another bacterial disease that infects honey bee larvae. However, unlike American foulbrood, European foulbrood does not create spores, so it's a bit easier to manage. European foulbrood causes the larvae to turn brown and twisted before worker bees cap its cell, so beekeepers (and honey bees) can see (and smell) infected larvae. Honey bees will remove these, taking them out of their cells and dumping them outside the hive. This is known as *hygienic behavior*, a trait that allows a honey bee colony to contain an infection or get rid of it completely. Beekeepers can purchase honey bee queens bred to produce offspring that provide hygienic behavior.

Both American foulbrood and European foulbrood have been around since the early 1900s. Both can be found anywhere honey bees are kept, and as such, they are pretty much a world-wide problem. Thankfully, apiary inspection programs can help beekeepers identify and control American and European foulbrood. These diseases are still a threat, but a manageable one, thanks to science and diligent, informed beekeepers.

Over time, two parasites eclipsed American and European foul-brood in terms of the havoc they wreaked on honey bee colonies.

One was tracheal mites. Too small to see with the naked eye, they make their home in the respiratory system of adult honey bees. Female tracheal mites enter the honey bee's *spiracles* (breathing holes in their bodies) and lay their eggs in the trachea. The eggs hatch, develop into adults, and eventually kill their host by clogging and/or damaging the trachea. The honey bee basically suffocates. The honey bee can also die from microorganisms entering the damaged trachea or from loss of hemolymph (blood). After mating with a male mite in the trachea, mature female tracheal mites will leave the host to find another honey bee to start the cycle again.

When tracheal mites first appeared in the United States in the mid-1980s, they hit the beekeeping industry hard, killing tens of thousands of honey bee colonies. The parasite spread around the world, but with time, treatment, and beekeeping management methods developed to control and prevent infestation, they gradually became less of a threat.

Then there are varroa mites. Varroa mites are reddish-brown, oval-shaped, pinhead-sized parasites. They latch onto honey bees with their tiny claws and suck out the bees' hemolymph with their piercing mouthparts. Although the mite is native to Asia, scientists believe varroa mites accidentally entered North America through Florida in the mid to late 1980s, hitchhikers in an illegal shipment of honey bee queens from South America. Honey bees in Asia (*Apis cerana*, a different species from *Apis mellifera*) have long been in contact with varroa mites and developed resistance; this is not the case for honey bees that had never been exposed to them. The mites spread across the United States and are still a serious concern around the world. Although varroa mites are small, they can be easily seen with the naked eye, so beekeepers can tell if their colonies are infested. Relative to their host, varroa mites are one of the largest parasites on the planet—it would be like humans carrying basketball-sized ticks on their bodies.

It's the adult female that latches onto the bee. After she feeds off her host, she looks for a place to reproduce, so she finds a larva inside a cell that has not yet been capped. She hides at the bottom of the cell, underneath the larva, until it has been capped with wax by a worker bee. Then she crawls onto the larva, lays eggs, and feeds of the larva's hemolymph. After the eggs hatch, the offspring feed off the larva's hemolymph too, and they defecate in the cell. As the larva develops into an adult honey bee, the varroa mites continue to feed off it and mature. The adult honey bee chews its way out of the cell, with the female varroa mites clinging to its body. (Male varroa mites stay in the cell for the rest of their life.) The honey bee carries these females into the colony, where they can jump onto other honey bees to feed. From there, the cycle continues with these females ultimately finding other larvae inside uncapped cells. If enough female varroa mites invade a brood cell, the larva might never develop into an adult bee and will die.

Although varroa mites do not directly kill adult honey bees, they weaken their immune system and shorten their lifespan, and they are a vector for a number of viruses, such as deformed wing virus, K-wing, acute paralysis, Kashmir bee virus, and others. Varroa mite numbers can explode in spring and summer, and they can quickly take over a honey bee colony. Previously strong colonies can quickly weaken and die.

Varroa mites are arguably the most devastating problem bee-keepers have faced to date, and have been called "the number one killer of honey bees on the planet." There is no silver bullet treatment to eradicate them from an infested colony or prevent a colony from becoming infected. However, research is ongoing, and there are currently a number of management practices that beekeepers can use to reduce the chance of losing their colonies and spreading the mites to others.

In addition to American foulbrood, European foulbrood, tracheal mites, and varroa mites, managed honey bees have faced other challenges throughout the decades: various bacterial and viral diseases, other parasites, fungal diseases, kleptoparasites such as small hive beetles, pesticides, and poor beekeeping management practices. These have all taken their toll. According to the National Agricultural Statistics Service (NASS), the number of honey-producing honey bee hives in the United States slipped from 5.9 million in 1947, to 4.5 million in 1980, and then to 2.4 million by 2008. Other countries such as Canada, Germany, France, Great Britain, and the Netherlands had reported elevated losses of managed honey bees over the years as well. Hayes and other experts knew that managed honey bee populations had been declining over the past half century. So, when honey bees began vanishing from apiaries in the early 21st century, it seemed like something new was hitting them that needed to be added to the list.

What was different this time was that the bees were literally gone. No dead bodies were found, so no autopsies could be performed. The honey bees had also left behind all their colony's eggs, larvae and immature bees, the queen, and all their stored honey. Just a small number of worker bees might be left behind, not nearly enough to sustain the colony and re-establish its previous numbers. No one knew why the abandoned contents of the hives had been left untouched by critters like robbing bees, wax moths, and small hive beetles that usually descend on such a bounty.

Large honey bee losses were not a new phenomenon in the history of beekeeping. As far back as the year 950, there were reports of a "great mortality of bees" in Ireland, which happened again in 992 and in 1443. One of the most famous honey bee catastrophes occurred in 1906, when most beekeepers on the Isle of Wight lost all of their colonies. In Colorado in 1891 and 1896, beekeepers witnessed the disappearance of large clusters of honey bees over a short period of time. They called the condition May disease. In 1903, in the Cache Valley of Utah, 2,000 colonies perished. The 1960s saw many reports of honey bee losses across Texas, Louisiana, and California. Australia and Mexico suffered high losses around 1975, and in 1995, Pennsylvania beekeepers lost 53 percent of their colonies. France had heavy losses between 1998 and 2000.

However, what was happening in 2006–2007 was unprecedented in the history of honey bee losses. Honey bees were disappearing suddenly across a number of states and countries, and many beekeepers lost most of their hives.

Hayes had contacted experts from Pennsylvania State University, the Pennsylvania Department of Agriculture, the Florida Department of Agriculture, and the US Department of Agriculture (USDA). "I remember sitting on my floor in my bedroom in Florida at the time while on this call with all these very smart people," he recalled. "We had no idea why this happened or would be happening, and so we decided to call it Colony Collapse Disorder [CCD]. So many things had happened in the beekeeping world, the industry of honey bees. We figured this would be just another one of them and it would be gone in 90 days and everybody would forget about it."

"Well, I guess it was a slow news day," Hayes continued, "and so media picked up on honey bees are dying. 'We don't know why, it's scary, the world is gonna collapse.'" Hayes said he had been in the beekeeping industry for a long time and it had never gotten that much attention from the media. "If it bleeds, it leads," said Hayes, "and so the thought that this organism that supports agriculture and the environment was sick, and we were losing 40 percent of the colonies a year . . ."

According to Hayes, the media failed to understand that if a beekeeper has two honey bee colonies and one dies, they can remove some of the honey bees from the remaining one (what is called "splitting" the colony), put them in an empty hive, and add a new queen. The two colonies can then build themselves up again. It may not be the best business paradigm, but that's how beekeepers have survived. And an unfortunate fact in the beekeeping world is that some honey bee colonies do not survive the winter.

Still, with so many beekeepers opening their hives to find that their honey bees were gone, Hayes and his team decided to dig deeper. They looked at data collected by the Apiary Inspectors of America and the USDA, who had surveyed beekeepers who managed anywhere from one to over 500 colonies. They found that over the winter of 2007–2008, 36 percent of all colonies died across the United States, about one million in total. This was an 11 percent increase from the previous year. Of the beekeepers surveyed, 38 percent felt their losses were not normal. On top of that, 60 percent of the wiped-out colonies across the United States contained no dead bees. The larger the apiary, the more likely the bees were gone. Also, apiaries that reported absent bees had significantly higher overall colony losses.

With so many beekeepers routinely moving their colonies around the country, it was difficult for Hayes and his team to tell whether colony losses were happening more in some regions of the United States than in others. They asked beekeepers what they thought was causing their losses, and their top five suspects were poor quality queens, starvation, mites, CCD, and weather. Hayes and his team were surprised by this, particularly because poor queens and starvation are manageable threats. Were beekeepers misdiagnosing the problem? Was there a need to change beekeeping practices and/or improve beekeeping education? Overall, the final report Hayes and his team published highlighted that beekeepers in the United States were experiencing increased honey bee losses, a significant number of these losses involved honey bees simply vanishing, but the causes were unclear. The fact that larger beekeeping operations were more likely to report absent bees suggested that whatever caused CCD could be contagious.

The United States government took notice. Managed honey bees are involved in the production of multitudes of food crops: almonds, apples, avocados, cherries, asparagus, broccoli, carrots, cauliflower, celery, cucumbers, onions, pumpkins, squash, sunflowers, and many others. At the time, the monetary value of honey bees as commercial pollinators in the United States was estimated to be about $15–$20 billion annually. Some researchers warned that the number of available commercial honey bee colonies was not keeping pace with the growing demand for pollination. More people living and being born in the United States meant more demand for food, which meant more bees were needed to pollinate crops. The country could not afford to lose them to CCD.

Between 2007 and 2010, the USDA poured about $33.5 million into funding for honey bee and CCD research. The agency set up a CCD Steering Committee that brought together representatives from a number of government agencies and academia. It established the Bee Research Laboratory in Beltsville, Maryland, to discover ways to improve the health of honey bees and help maintain an adequate supply of commercial honey bee colonies to pollinate crops. After years of intense research, the consensus is that there is no one cause behind CCD and it is likely a result of a number of factors working in combination or synergistically. The possibilities include:

- Sublethal effects of pesticides on honey bee health and behavior;
- *Nosema ceranae*: a single-celled parasite that invades the honey bee's gut, causing diarrhea and lessening the bee's ability to absorb nutrients when it eats;
- High levels of varroa mite infestation;
- Disease, such as acute bee paralysis virus and deformed wing virus;
- High levels of bacteria or fungi;

- Poor nutrition due to apiary overcrowding;
- Pollination of crops with low nutritional value;
- Scarcity of pollen or nectar, which can be a result of climate change;
- Exposure to limited or contaminated water supplies; and
- Stress from shipment between crop sites.

Each of these is bad on its own; a combination of two or more of these stressors can weaken a colony enough to make it collapse.

In Hayes's opinion, the term CCD has been overused. Honey bee disappearances appear to be a reaction of colonies that are beyond their tipping point. "When the workers get sick," Hayes explained, "they will individually leave the hive and not come back because they do not want to expose their sisters to whatever is making them sick. They were leaving to sacrifice themselves, if you will." Hayes relates this to how humans have coped with Covid-19: "Like you, if you have Covid, you don't wanna go to your mom and dad's house."

All of the doom-and-gloom media attention honey bees and CCD received made Hayes roll his eyes. However, one good outcome of all that attention was that the government was finally providing money to support previously ignored research on the beekeeping industry and on honey bee health in general.

But what about other species of bees that were disappearing?

Historically, Franklin's bumble bee was found in Douglas, Jackson, and Josephine counties in southern Oregon, and Siskiyou and Trinity counties in northern California. This range—the smallest of any bumble bee in North America and possibly the world—forms an oval about 306 kilometers north to south and 113 kilometers east to west. Franklin's bumble bee shared this area with a number of other bumble bee species. Many of the sites where Franklin's bumble bee used to be seen were parks, meadows, and trails that had been left relatively undisturbed over the years. In this habitat, Franklin's and other bumble bees could feed from flowers, find a nesting site, and hibernate for the winter. Pesticides do not seem to have been applied to the areas. So why did Franklin's bumble bee vanish, and why did it vanish so quickly? Thorp's theory is disease.

Earlier I mentioned the commercial bumble bee industry, where Europe breeds the buff-tailed bumble bee and North America breeds the common eastern bumble bee.

For a while, *Bombus occidentalis*, the western bumble bee, was commercially available as well. That's where the story of the fall of Franklin's bumble bee is thought to have begun.

In the early days of the commercial bumble bee industry, a number of queens of both the common eastern bumble bee and the western bumble bee were shipped from the United States to rearing facilities in Belgium. These facilities were thought to have also reared buff-tailed bumble bees. Once colonies of the common eastern bumble bee and western bumble bee were ready, they were shipped from Belgium back to the United States—bringing with them *Nosema bombi*, a fungus-type parasite that infects bumble bees' excretory tubules, fat body, nerve cells, and sometimes their tracheae. The effects can be mild to severe, and infected colonies can appear asymptomatic and healthy. *Nosema bombi* likely jumped from colonies of the buff-tailed bumble bee at the rearing facilities to colonies of the western bumble bee.

Disaster ensued. A major outbreak of *Nosema bombi* occurred in commercial western bumble bees in North America. It was so bad that colonies had to be destroyed and production of the western bumble bee completely stopped. Then the population of Franklin's bumble bee took a nosedive. Wild populations of the western bumble bee also plummeted. Thorp's theory was that commercial western bumble bees that had escaped from their greenhouses shared flowers with wild populations of the western bumble bee and Franklin's bumble bee, and spread *Nosema bombi* to them, a phenomenon often referred to as pathogen spillover. It is possible that other species of bumble bees still seen where Franklin's once was were not as affected by the fungal parasite. For unknown reasons, Franklin's bumble bee might have been particularly susceptible.

When it comes to the rusty-patched bumble bee, its decline also happened to coincide with the growing use of commercial bumble bees for greenhouse pollination in the late 1990s and early 2000s. Historically, the rusty-patched bumble bee could be found across a broad range, from the eastern United States and upper Midwest, north to Maine in the United States and southern Quebec and Ontario in Canada, south to the northeast corner of Georgia, and as far west as North and South Dakota. Studies have shown that the rusty-patched bumble bee is no longer found in 70 to 87 percent of this range. Where the rusty-patched does occur, its numbers have shrunk by up to 95 percent. So, when Colla found only one rusty-patched bumble bee where it used to be commonly seen, her experience wasn't unique. In a number of areas, the bee has become extremely rare.

Similar to the case of Franklin's bumble bee, parasites spread by commercial bumble bees might have led to the decline of the rusty-patched bumble bee. Again, *Nosema bombi* is a major suspect, but commercial bumble bee colonies in North America have tested positive for at least seven other pathogens. These could have led to the decline of wild bumble bee populations in three ways. One is that commercial colonies could tolerate a lot more of a naturally occurring pathogen than is found in the wild, so bumble bees like the rusty-patched might not be able to cope when exposed to such high levels. Second, the conditions created by the commercial production of bumble bees can provide a breeding ground for more virulent strains of pathogens to evolve. In this case, the introduced pathogen is much more powerful than wild bumble bees' immune systems can handle. Finally, introducing a commercial colony to an area where that species of bumble bee was not naturally found, particularly in the case where colonies are shipped across the globe, can expose wild bumble bee populations to pathogens they have never encountered before.

Was Thorp's theory about commercial bumble bees escaping the possible answer? They certainly can slip out of a building through open windows, ventilation systems, and even cracks. One way to tell whether commercial bumble bees are escaping from greenhouses is to examine the pollen loads that they bring back to their nest. If the pollen they carry differs from the pollen of the greenhouse plants, then you can be sure that the bees ventured outside. This is exactly what a team of scientists at Simon Fraser University found. Over eight months, in different greenhouses in southwestern British Columbia, they caught greenhouse bumble bees returning to their nest after a foraging trip. They gently scraped the pollen off the bees' legs and examined it under a microscope. What they found was quite surprising. At one of the greenhouses, bumble bees had brought back 95 percent tomato pollen; 5 percent of their pollen was from plants outside. At another greenhouse, as much as 73 percent of the pollen samples were from non-tomato flowers such as blackberry, raspberry, thimbleberry, salmonberry, dandelion, thistle, foxglove, fireweed, buttercup, and pink spirea. All these plant species bloomed in the area surrounding the greenhouse. Notably, throughout the course of the experiment, there was never a lack of tomato flowers in the greenhouses. The commercial bumble bees were sneaking out, providing their colony with more variety in their diet.

Once a commercial bumble bee infected with a pathogen escapes from a greenhouse, it can spread the pathogen to wild bumble bees when they share flowers, in the same way humans can pass germs to each other by contaminating surfaces such as doorknobs. Contaminated nectar can also spread illness to wild bees, the way humans sharing straws or drinking glasses can infect each other.

In the summers of 2004 and 2005, Colla and a group of her scientist colleagues set out to determine whether pathogen spillover was indeed possible from commercial bumble bee colonies in greenhouses to wild bumble bee populations. If pathogen spillover was happening, then the wild species of bumble bees they caught near greenhouses would show higher instances of pathogen infection. And in fact, across six different locations in southern Ontario, they identified the intestinal parasite *Crithidia bombi* in a number of wild species of bumble bees captured near greenhouses; but it was absent in bumble bees caught farther away. *Nosema bombi* was three times more likely to be found in wild species of bumble bees caught near greenhouses.

Wild bumble bees have also been known to "drift" to other nests, either to steal stored nectar or because they somehow become disoriented and end up in the wrong home. If these "drifters" visit an infected nest, then return to their own nest, this is another way that pathogens can be spread. One team of scientists in Newfoundland, Canada, found a number of drifters inside boxes of commercial common eastern bumble bee colonies that had been placed in open fields of blueberries and cranberries. These drifters included several different wild species: the tricolored bumble bee, the half-black bumble bee, the frigid bumble bee, and the yellow-banded bumble bee. When the scientists screened those commercial colonies for seven possible diseases, they tested positive for three. The potential was there for infected commercial bumble bees to pass these on to wild bee populations through these drifters.

Unlike Franklin's bumble bee, the rusty-patched bumble bee also faced drastic changes to its habitat. The grassland, meadow, and woodland ecosystems where the rusty-patched bumble bee lived have been drastically fragmented or lost, in some cases to agriculture or urban development, in others, to invasive species. The fires that once kept land open were prevented. As a result, their remaining habitat has become extremely patchy. This has major consequences, because bumble bees require three different types of habitat, located near each other: a place to nest, a place that provides food (pollen and nectar) from early spring to early fall, and a place for queens to hibernate over the winter. Historically, grasslands, meadows, and woodlands provided all three.

These days, nesting areas are more limited. Queens and worker bees may have to travel farther to find food, and finding undisturbed soil in which queens can hibernate can be challenging. The rusty-patched bumble bee is also a short-tongued species, so it cannot necessarily collect food from just any type of flower. They simply can't reach the nectar of some flower species. In addition, patchy habitat can lead to inbreeding. Small pockets of habitat end up isolating bumble bee colonies that once interacted from each other. If a stretch of habitat can only sustain a small number of colonies, male bumble bees end up mating with queens that are too closely related to them. New generations of bumble bees will have less genetic variety, reducing the bees' ability to ward off stress and infections. Male bumble bees may also become sterile. It would not take long for the bumble bee population to subsequently crash.

We may never confirm the exact details of why and how the rusty-patched bumble bee suffered such a significant decline. However, it is clear that over time the species has faced a perfect storm of stressors. Humans have drastically altered its habitat, hampering its ability to nest, gather adequate nutrition, reproduce, avoid inbreeding, and survive winter. Such ongoing stress can lead to weakened immune systems, and death, and ultimately a decline in the species' population. Add an influx of disease from commercial colonies, and the rusty-patched bumble bee may no longer be able to cope.

When it comes to the struggles of the Hawaiian yellow-faced bees, as humans constructed buildings, roads, and other infrastructure on the Hawaiian islands, available habitat for the bees shrunk. Habitat was also transformed by introduced non-native plants that spread and overwhelmed native species that provided food, nesting sites, and shelter for Hawaiian yellow-faced bees. But the biggest threat? Quite possibly ants.

All ant species found in Hawaii have been introduced. Some scientists believe that Hawaiian yellow-faced bees probably nested in the ground a lot more often in the past, but high densities of ants have forced them to find alternative homes in plant stems, washed-up coral, under rocks, or in other tiny, hidden cavities. Ant colonies may find and establish themselves in nesting sites that would have once been occupied by Hawaiian yellow-faced bees, or they may physically evict bees that are already there. Scientists observed many cases of ants forming lines into nests of Hawaiian yellow-faced bees and removing pollen, eggs, and pupae to make room for their own colony to take over. Ants on the Hawaiian Islands also compete with the bees for floral nectar, and Hawaiian yellow-faced bees have been known to avoid certain flowers if ants are present. Other introduced insects like wasps and managed honey bees can put added stress on Hawaiian yellow-faced bees by monopolizing even more resources and potentially introducing disease.

Some Hawaiian yellow-faced bees live in coastal areas just above the high tide line. In February 2016, all their nests in the peninsula at Turtle Bay, Oahu, were completely washed away by a giant swell. Human-accelerated climate change triggering a rise in sea levels, combined with an increase in storms and anomalous swell events, could be catastrophic for the survival of many of Hawaii's only native bee species.

From what is currently known, the disappearance of millions of honey bees in the early 2000s and the decline of various species of wild bees were likely caused by a tangled web of factors working together. In some cases, one factor might have been more influential than others. This was probably the case with pathogen spillover and Franklin's bumble bee, since disease is the culprit most likely to have caused such a drastic population collapse in such a short amount of time.

Bees have been—and still are—facing a multitude of stressors in a complex world. The stress that managed honey bees are facing can be different from the stress that wild bee populations are coping with. To add to the complexity, in both wild and commercially reared populations, some species of bees appear more resilient than others, for reasons we have yet to fully understand. As a result, we cannot expect a single silver bullet to "save" all of them. Certainly not urban beekeeping. It adds more honey bees or replaces vanished ones rather than addressing honey bee (and other bee) diseases and related health issues. It's kind of like handing a Band-Aid to someone with serious internal bleeding.

One more thought. There is now what I believe to be convincing evidence that bees have a mental life and quite possibly experience pain, emotions, and consciousness. Whole books have been written on this subject recently, including Lars Chittka's *The Mind of a Bee* and Stephen Buchmann's *What a Bee Knows*. (They are both excellent and I highly recommend them to readers who wish to know more.) Given this, I can't help but wonder what it must be like for a bee to experience any of the diseases described in this chapter. What is it like for them to feel sick? What is it like for them to witness their colony succumbing to illness? What is it like for them to discover that large numbers of larvae in their hive are dying? What is it like to be a bee in a colony that is so stressed and/or ill that it decides to flee, leaving everything behind?

Are our actions responsible for the suffering they are quite possibly experiencing?

Chapter Five

P RETEND WE HAVE transported our naked Spanish friend from ancient times to the present. No doubt he would be blown away by the millions of ways the world has changed since his time. So many people! Buildings! Planes, trains, and automobiles! If he were to look out into the landscape where he lived long ago, he would see that it has been cleared for fields of crops and orchards. He would also see that people have placed wooden beehives near many of these. These beehives contain the descendants of the honey bees he stole honey from way atop that precarious rock wall thousands of years ago. We could show him how we can harvest honey from these hives without risking our lives like he once did. We could also explain that beehives are often placed near crop fields because honey bees help produce the food we harvest.

However stunned he might be by our world, he might notice some things that remained the same. If we were to spend some time with our ancient friend watching the flowers in a Spanish melon field, as well as those honey bees, we might see upwards of 40 different species of bees including various species of sweat bees. If we were to take him to an almond orchard, we might see 20 different bees visiting the blooms, including many solitary, ground-nesting bees from the *Andrenidae* family. Although honey bees would be one out of the many different species we'd see, they certainly wouldn't be alone.

The odd thing? To him, they might be familiar; to us, they are too commonly invisible, despite the key role they play in our very survival.

What we saw with our friend in those fields, isn't just something particular to Spain. Scientists Sabrina Rondeau, Susan Willis Chan, and Alana Pindar looked at published records of bees at 23 different crops across Canada and the United States. They tallied more than 700 species of wild bees visiting those plants. They listed the number of wild bee species according to crop type as follows:

- Blueberry: 131 species
- Apple: 83 species
- Strawberry: 82 species
- Cherry: 68 species
- Watermelon: 62 species
- Cranberry: 61 species.

More than half of these species were ground-nesting bees. Of all of these wild bees, the most common species were *Lasioglossum* (sweat bees) and *Andrena* (mining bees).

Looking at farms in Virginia over the course of a single summer, N. L. Adamson, T. H. Roulston, R. D. Fell, and D. E Mullins discovered that this diversity acts as a sort of "insurance policy," guaranteeing that there are always bees around to pollinate. Looking at farms of various sizes that grew apples, blueberries, raspberries, blackberries, summer squash, winter squash, pumpkins, cucumbers, cantaloupes, and watermelon, they observed that the life cycles of wild bees often result in different species working different "shifts" over the course of the spring and summer to pollinate plants. For instance, there are lots of mining bees and mason bees in June, but they didn't see bumble bees until later in summer. So, despite the natural fluctuations in abundance of different bee species over a growing season, there is always someone around to pollinate the plants.

In both these cases, managed honey bees or commercial bumble bees are used to pollinate these crops. But interestingly, compared to crops pollinated only by wild bees, this does not always result in better or higher fruit set (the development of flower to fruit). This seems to be the case even in small, isolated areas such as islands. In isolated Newfoundland off the mainland of Canada, observers there would see much less wild bee diversity when compared to other areas of North America, but they would still notice wild bees frequenting the island's cranberry and blueberry crops. For blueberries, Newfoundland's native Carolina miner bee (*Andrena carolina*), the Quebec sweat bee (*Lasioglossum (Evylaeus) quebecense*), and the half-black bumble bee (*Bombus vagans bolsteri*), seemed to get the job done. For cranberries, the orange-belted or tricolored bumble bee (*Bombus ternarius*) and metallic sweat bee (*Lasioglossum (Dialictus)*) species are common and efficient pollinators. Wherever you go, wild bee species might be the ones doing the heavy lifting in terms of pollinating food crops.

People have assumed that adding managed honey bee hives to their fields and orchards would ensure or improve good yields and high quality fruit. It's part of what might be called the "foundational" argument for the inherent value of honey bees. But for some crops in some regions, this might not always be the case. For instance, sweet cherry orchards in Belgium are often supplemented with managed honey bees, and they are indeed the most frequent visitors to the cherry blossoms. However, it was the orchards that also had a large wild bee community that produced the best yields. Four times higher, in fact. It seems that *how* bees visit cherry blossoms, not how often, determines the quality of their yield. Wild bees almost always make contact with a flower's sexual parts during a visit, thereby ensuring pollination. Managed honey bees, on the other hand, often visit cherry flowers from the side, miss contact with those parts, and fail to pollinate the flower.

Managed honey bees might also not be that helpful for strawberry plants. When fine mesh bags were used to control which type of bee visited strawberry flowers at a commercial farm in Québec, Canada, the best fruit came from flowers that were visited by wild bees only. Managed honey bees actually resulted in *lower* fruit yields compared to the yield that would have been produced by relying solely on the wild bee community.

What were the wild bee species that helped to produce the best strawberries at the farm? *Lasioglossum* and *Augochlorella* were the superstars. These small ground-nesting sweat bees, usually metallic green in color, measure about five to seven millimetres in length. What is particularly interesting is that these bees were seen visiting the strawberry flowers much less frequently than honey bees, yet they managed to get the job done. Similar to what we saw earlier with sweet cherries, the key to the wild bees' success could be their foraging technique while on the strawberry flower. They are able to forage for pollen and nectar without shifting the anthers. Larger bees (10 millimeters in length or greater) often end up bending and moving the anthers toward the stigmas while searching for nectar. This could result in the flower self-pollinating, which for the variety of strawberries that were at the farm, produces smaller fruit than flowers that are cross-pollinated.

It is also possible that how the strawberry plants were planted was a factor. Honey bees have been known to work their way down a row when foraging. If a row contains the same variety of strawberry, smaller fruit could result from lack of cross-pollination with other varieties. In addition, with managed hives each containing tens of thousands of honey bees, these can saturate strawberry stigmas with pollen. The potential result is poor yields for the farmer.

Like strawberries, apples are a crop that relies on cross-pollination between plants in order to produce good quality fruit. A number of wild pollinators, such as bumble bees, solitary bees, and hoverflies, are good at this.

One research team in Argentina decided to find out just how important one of these wild bees species is for producing good apple yields there. The bee in question was *Bombus pauloensis* (formerly known as *Bombus atratus*), and also known as the black bumble bee, Paulista bumble bee, or Mangangá negro. (For simplicity I'll refer to this bee as Paulista.) This particular bee almost looks like a tiny, buzzing black bear. Paulista is also a good pollinator of other crops, such as tomatoes, strawberries, and peppers. (Sadly, in recent years, many wild pollinators have disappeared entirely from Argentinean orchards, likely because of intensive crop management and use of pesticides. One of these missing pollinators is Paulista. However, it has been bred in captivity and is now commercially available in Argentina.)

Managed honey bees were present across eight orchards, and the researchers added commercial colonies of Paulista in four of them. In orchards with both managed honey bees and Paulista, the researchers saw honey bees visiting the apple flowers almost twice as often as Paulistas. However, orchards with Paulista present produced four times more apples per tree. The number of developed apple seeds was also higher. Similar to what we saw earlier with cherries and strawberries, the presence of native bees in cropland results in better pollination and better crop yields, despite the large numbers of managed honey bees that might also be seen on the flowers.

Do wild bees contribute to the pollination of apples around the world, like they do in Argentina? The consensus is yes. One research team examined apple pollination across forty-six commercial orchards in Belgium, France, Morocco, the Netherlands, Spain, and the United Kingdom. Overall, the high numbers of managed honey bees seen visiting apple trees and the number of managed honey bee hives in apple orchards had no significant impact on apple yields. Wild bees were the ones successfully pollinating the apple trees. Based on their findings, the research team recommended supporting wild bee diversity rather than relying on managed honey bees for commercial apple production. A recent meta-analysis of apple pollination research around the world came to a similar conclusion: Wild pollinators are important for apple pollination, and supporting these wild pollinator communities where they naturally exist is the preferred strategy for ensuring successful apple yields. In some cases, managed honey bees might not be necessary and might even be detrimental to crop yield. Relying on wild bees also has the added benefits of supporting pollination of local wild plants and conserving biodiversity.

Compared to what is known about honey bees and bumble bees, we know much less about the individual players in this silent workforce. However, as we learn more, it is clear that these bees are important. And incredible, each in its own way. Let's now meet two very cool, lesser-known bees: *Halictus ligatus* and the hoary squash bee.

Halictus ligatus is a type of sweat bee that is a pollinator of at least fourteen different crops across North America, including cotton, soybean, sunflower, apples, blueberry, and peppers. Also known as the ligated furrow bee, this little sweat bee has quite a broad range, stretching from southern Canada to Colombia and Trinidad. It has been particularly well studied in Ontario, Canada, where it is commonly found in the summer, especially in July. It is a medium-sized bee, around seven to ten millimeters long, and black with yellow bands on its abdomen. Unlike honey bees and bumble bees, *Halictus ligatus* is not very hairy. However, it is unique looking. "If you can get close enough to a flower, you can tell the species because they have a rather big head for their body," scientist Sandra Rehan told me. She is an expert on social insect genomics and pollinator health, and she knows *Halictus ligatus* very well. "They have this really cute little spine off their cheek that distinguishes the species. It's a cute little bee."

True to the name "sweat bee," Rehan experienced *Halictus ligatus* using her as a salt lick during her field work. "I spent summers excavating colonies and so because I was like the only salt source around, they sometimes would land on me. Especially if I was sweating in the field digging up bee nests. They do land on you and drink your sweat."

Halictus ligatus is a ground-nesting bee that lives in small colonies. In the spring, mated queens that hibernated over winter emerge from under the ground and establish their nests by burrowing into soil that has good drainage and a certain amount of sun exposure. Queens often choose nesting sites that are in bare, exposed areas or that feature short vegetation. They may even return to previously occupied nests and refurbish them. The nests have perfectly round entrance holes about the diameter of a pencil. "They have this hovering flight they do over the nest entrance," Rehan told me. "So, you can see them over the ground." Queens tend to build their nest nearby each other in aggregations. "To go out one day and find a nest is actually hard, but if you do find an aggregation, they're often one hundred or more in a patch."

Each queen's nest is a tunnel about 15 centimeters deep with about six or seven cells branching off it. The queen lines each cell with a wax-like material, and in each cell, she lays an egg. She closes the entrance to her nest with earth and stays inside until her daughters hatch and mature into adult worker bees. These workers leave the nest to forage for pollen and nectar, and they return with the food and help the queen look after her brood. The queen stays home for the rest of the summer, constructing cells and laying eggs. The tunnels in the nest may branch out and extend deeper over time. Usually there are fewer than ten worker bees, and by late summer, it is males and new queens that emerge from the eggs the queen lays. These males and new queens leave the nest, mate, and the male bees die off. The original queen and any remaining worker bees die off as well, and the new queens go underground to hibernate for the winter. In spring, they dig their way out, and the cycle begins again.

However, female sweat bees have shown some interesting flexibility in their behavior. Occasionally, queen bees will "share" a nest after they emerge from hibernation in the spring. One queen becomes dominant, laying the eggs and guarding the nest, whereas the other bee behaves like a worker bee, foraging for nectar and pollen. (Nest guarding involves the queen sticking her large head at the nest entrance and biting at any intruders, or bending her body into a C-shape with the end of her abdomen facing the entrance, ready to sting.)

Adult female *Halictus ligatus* bees can also switch their role from queen to worker, and from worker to queen. A small number of male bees can appear early in the colony cycle in spring, and if the queen dies, a worker bee can take over as the new queen. This worker bee would have mated with an early spring male bee during one of her foraging trips. After the worker bee takes over, it is business as usual: like her original queen would have done, she lays eggs that hatch as female workers, she stays in the nest while the workers forage, and later in the summer she will lay eggs that produce new queens and males. It is also possible for some worker bees to hibernate over the winter and become nest foundresses, as they are known, in the spring.

Rehan witnessed yet another level of flexibility in this bee. One summer in early July, she was studying ground-nesting bees and was looking for nest aggregations on the Brock University campus in St. Catharines, Ontario. "Near a riverbed there was a big dirt pile, just kind of a leftover from, I guess, landscaping or something on campus," she told me. "It was kind of off the beaten trail in the backwoods of Brock University campus. So not in a common area, not a public area which makes it easier, because most ground-nesting bee aggregations don't last very long because something will drive over it or plow it. They're very easily disturbed." Rehan was fortunate to find two nest aggregations that summer. Most were *Halictus ligatus* nests, but there were also *Halictus confusus* (also known as the confusing furrow bee or southern bronze furrow bee), *Halictus rubicundus* (also known as the orange-legged furrow bee), and some *Lasioglossum* bees (another genus of sweat bees). "At one point there was a turtle. A turtle made a nest nearby, so it was pretty cute."

One morning when she returned, the area was a complete mess. The bee nest entrances were gone, as if a small tornado had swept across the soil. But it wasn't a tornado; it was black *Astata* wasps. They had likely overwintered deep within the soil, and they had finally emerged. The wasps are about twice the size of *Halictus ligatus* bees, and by digging out of the soil, they substantially disrupted the landscape. They were also digging their own nests, which further destroyed the *Halictus ligatus* nests. As a witness to the scene, Rehan said the wasps "created mass chasms and just chaos." She saw many *Halictus ligatus* females back from foraging, hovering over the soil, searching for their nest entrances. "And so this bee that's normally eusocial with queen and workers, all of a sudden these workers were all kind of rendered homeless," Rehan told me. (Eusocial means living in a group with overlapping generations, cooperative brood care, and only one or few individuals reproduce and the rest are sterile.)

Then, later that summer, Rehan saw something quite remarkable. Four of the orphaned *Halictus ligatus* worker bees had started nests of their own around the perimeter of the new wasp nest aggregation. They switched their role from worker bee to queen, establishing their own nests like queens do in the spring. One of the four bees was likely a queen who survived her nest destruction; she was larger than the other three bees and showed a level of wing wear typical for older queens. Other species of sweat bees had been observed founding nests in the summer, but not *Halictus ligatus*. "Everyone had reported over the years that this bee was obligately eusocial," said Rehan. That means "queens make workers, workers work and help produce gynes [queen bees] in the fall. This showed that the bees are actually much more adaptable given unpredictable environments, given nest usurpation or whatever you want to call that. They were able to actually go out and dig their own tunnel and they just kept foraging and doing things solitarily, which had never really been reported before. You'd think they'd just be rendered homeless and die, but nope. They rose to the occasion, I guess."

Two years ago, my husband installed several large raised garden beds in our backyard. In one of them he planted zucchini and buttercup squash. Mid-morning one summer day, I stood admiring the bright yellow blooms when I happened to notice a cluster of small insects inside one of the flowers. I peered closer and saw what looked like three little honey bees, side-by-side, clinging to the pistil of the flower (the tall stalk in the center). They were facing up, looking back at me. They didn't move and I didn't hear them buzz. They seemed to be resting, just hanging out together in the squash flower.

Later in the afternoon, I saw that the flower had closed. I gently parted the puckered, wilted, twisted tips of the delicate petals for a look inside. There, inside the flower in a snuggly cluster, was a three-bee slumber party. Thinking that they probably didn't appreciate my disturbing them, I let go of the petals, sealing the bees back up inside their cozy little sleeping chamber.

Curious, I moved on to the next closed squash flower on the vine. I carefully parted its petals, peeked inside, and saw two similar little sleepy bees. Like the others, they were curled next to each other near the bottom of the flower.

What I was seeing were not honey bees, but hoary squash bees (*Xenoglossa* (*Peponapis*) *pruinosa*). I was witnessing a deeply established, intimate relationship between a special type of bee and special types of flowers.

At first glance hoary squash bees do look like honey bees. They are approximately the same size and body shape, and they have a similar amount of hair concentrated on their thorax. However, hoary squash bees (they are also sometimes called eastern cucurbit bees or eastern squash bees) can be distinguished by the gray and bright white stripes on their abdomen (hoary means gray or grayish). Compared to honey bees, their flight is much faster and darting, almost like a fly's. While female honey bees carry pollen in compact, moist pellets on their smooth back legs, when female hoary squash bees gather pollen, it looks more like a cloud of grains attached to their much hairier back legs.

Hoary squash bees behave quite differently from honey bees, too. A solitary species, hoary squash bees don't live in a colony of tens of thousands of family members. Although small groups of males and unmated female hoary squash bees do snooze together inside closed squash flowers, they otherwise live on their own.

After a female hoary squash bee mates, she needs a place to lay her eggs. Each female hoary squash bee excavates her own nest in the ground. She first locates bare soil in open areas near squash plants, or a lawn where the grass is kept short, and then she digs a vertical tunnel about six or seven inches deep. She hauls all the soil from the excavation of the tunnel up to the surface where it accumulates in a heap around the entrance. At the bottom of the vertical tunnel she excavates a horizontal shaft that ends in a nest cell. She hauls all that soil up to the surface, too. The cell will eventually be the birthplace of her offspring, so, as a good mother, she needs to prepare the space for her baby to thrive. She lines the cell with a waterproof, cellophane-like secretion. She fills about a third of the cell with pollen and nectar, which requires, on average, about twenty foraging trips. Once the cell is well-stocked with food, the female hoary squash bee lays her egg on top of the pile of provisions. Her last gesture as a mother is to seal the cell. The egg is thus buried underground, tucked away nice and cozy within a waterproof cell.

The mother hoary squash bee does not wait around for her egg to hatch and develop into an adult bee. She has more eggs to lay and thus more work to do. She repeats the process. She moves upwards in the vertical tunnel and digs about three or four more horizontal tunnels. She uses the excavated soil to back fill the vertical tunnel. Like before, each horizontal tunnel she makes ends in a terminal cell, and she waterproofs each one and supplies them with pollen and nectar. She lays an egg in each, seals them up, and leaves them to hatch and develop on their own.

At some point the mother hoary squash bee rests. Not rest within wilted, closed squash flowers, however, but in an antechamber that she excavates at the top of the vertical tunnel she had dug, above the nest cells and just under the surface of the ground. After digging this, her work is not done, however; she has more eggs to lay and so she will dig more nests throughout the summer. These nests can be identified by the mound of loose soil she leaves at their entrance. They look like tiny volcanoes and can be mistaken for anthills, each with an entrance that is about a pencil's width in diameter. That first nest that the mother hoary squash bee dug was likely near the nest from where she herself emerged as an adult bee. As long as there are squash plants nearby, squash bees will often nest in the same location year after year.

Although hoary squash bees are a solitary species, the females will often build their nests next to each other, creating what are called nesting aggregations. Some aggregations can include hundreds of nests. As the females zip in and out of their nests in these aggregations, foraging for pollen and nectar to feed their future off-spring, the result can look like a swarm of honey bees. However, hoary squash bees are a gentle sort. They do not actively defend their nests like honey and bumble bees, and pretty much ignore any human or other creature walking among their nests. However, they may sting in defense if their nest entrance is blocked or if they are handled.

Returning to their lifecycle, after the mother hoary squash bee lays her eggs in her homemade underground nursery, she leaves them to develop on their own. Each egg hatches into a tiny, worm-like larva, and it feasts on the pollen and nectar that its mother provided for it. The larva grows, develops into a pupa, and emerges as an adult hoary squash bee the following year. There is one generation per year. If the newly emerged hoary squash bee is a male, he will spend most of his life hanging out by squash plants, "cruising" if you will, for mates. As he does, he inadvertently picks up pollen grains, and ends up transferring them between flowers. In this way, male hoary squash bees are very effective pollinators without really meaning to be. After a morning of wandering, mating, and snacking on nectar, the male bee will take shelter in a squash flower. The flower will close in late morning or around noon, depending on the outdoor temperature. This allows the male bee inside to snooze safely all afternoon and night. Early the next morning, he will chew through the wilted petals or push his way out, resuming his search for unmated females (and inadvertently pollinating squash flowers). Females mate with a single male, but males might mate with multiple females.

Newly emerged female hoary squash bees may hang out and snooze with the males in squash flowers, but once they mate, they begin a solitary life of building and provisioning nests and laying eggs. Female hoary squash bees also inadvertently help squash plants: although they collect the flowers' pollen to feed their young, some pollen grains that stick to the hoary squash bees' body are transferred between flowers. Considering each female hoary squash bee needs an average of twenty flower visits to collect enough pollen for just one nest cell, that's a lot of squash flower pollination.

If you spend some time in a squash field or in a garden with squash plants, you'll likely see a variety of critters visiting the squash flowers. The large, yellow, showy blooms attract a variety of pollinators like beetles, ants, flies, and in some areas, over thirty species of bees. They come to the flowers for the nectar, which is rich in sugars and thus a great source of energy. The scent of the flowers attracts pollinators, too. Scientists who studied squash plant visitors in the northeastern United States found that most flower visits were by honey bees, bumble bees, sweat bees, the two-spotted long-horned bee, and of course, squash bees. (Besides the hoary squash bee that I saw in our garden, there are roughly twenty other species of squash bee across North America.)

Interestingly, although squash plants produce pollen that is rich in protein and fats—nutrients that are essential for the development of larvae into adult bees—squash bees appear to be the only type of bee that actively collects it to bring it back to their nest. When pollen sticks to a female bee's body, she scrapes it or grooms herself so that the pollen collects on her back legs. With more and more flower visits, the amount of pollen on her back legs accumulates quite significantly, looking like what I've heard endearingly described as "pollen pants." Scientists observing bees at Pennsylvania squash farms watched as honey bees and bumble bees landed on a leaf after taking a drink of nectar from a squash flower, and scraped off any pollen grains that remained on their body. Why did these bees not want to bring squash pollen back to their nest to feed their young?

To find out, scientist Kristen Brochu and her team fed cucurbit pollen to the larvae of small colonies of common eastern bumble bees. (Squash, zucchini, pumpkin, and gourd plants are in the genus *Cucurbita*, and are collectively referred to as cucurbits in the scientific literature. I'll refer to these plants and fruit as cucurbits going forward.) The team then saw the adult bees remove the larvae from their brood cells and dump them outside of the nest: a sign that adult bees are stressed and/or their larvae are sick. No larvae were reared to adulthood. Up close, pollen grains of cucurbit flowers look large, spiky, and sticky. When they examined the guts of the bumble bee larvae that were fed cucurbit pollen, they were quite swollen with dark patches. But Brochu and her team discovered that it wasn't just the physical aspects of the cucurbit pollen that played a part. The chemical composition and lower nutritional value of the pollen also contributed to bumble bee larva mortality. I can't help but wonder that for bumble bees (and perhaps honey bees and other bees species as well), eating cucurbit pollen is kind of like eating balls of tinfoil.

Strangely enough, however, squash bees collect cucurbit pollen pretty much exclusively and cucurbit pollen allows their larvae to mature into healthy adult bees. Somehow their digestive system can cope with the large size, spikiness, and chemical contents of the pollen, and indeed the larvae have come to rely on it to grow. This need for cucurbit pollen, and the fact that other bees reject it, speaks to the intimate relationship between squash bees and cucurbit plants that has endured for millennia.

The lives of squash bees and cucurbit plants are synchronized, both seasonally and on a daily basis. The first adult squash bees of the season emerge in the month when flowers appear on cucurbit plants and these bees remain active throughout the blooming period. In California, Utah, and Mississippi, hoary squash bees are active in cucurbit crops beginning in May and over the two-month blooming period. In the northeastern United States, they are active in July. In Ontario at the northern edge of their range, between mid-July and the end of August. Each day during the cucurbit's blooming period, hoary squash bees are early risers, awake and buzzing about first thing in the morning, sometimes before sunrise, when the cucurbit flowers begin to open and before any other bee species have begun to forage. This way, they are the first bees to take advantage of the bounty of nectar and pollen that cucurbit flowers have to offer.

Male hoary squash bees also take advantage of the synchronicity between their species' daily activity and the opening of cucurbit flowers, by using the flowers as a place to mate. Hoary squash bees often mate in cucurbit flowers, with male bees pouncing on females when the females are trying to collect pollen or nectar, or when they are resting. Sometimes more than one male will try to mate with a female. This can result in a fumbling ball of bees that has been seen on occasion tumbling out of the flower.

The love story between cucurbit plants and squash bees can be traced back at least 6.5 million years. Squash bees are thought to have originated in northwest Mexico at that time (cucurbit plants had probably already existed for about 4.5 million years before that). As far as cucurbit plants were concerned, squash bees were a welcome addition to the ecosystem because the plants had a problem: they produce separate male and female flowers. And for the plants to reproduce, pollen has to travel from a male one to a female one. Because the pollen grains are large and sticky, wind pollination wouldn't work. And who better than squash bees that, as we have seen, benefit from cucurbit pollen? A win-win situation.

Scientists are piecing together the details of the history of the cucurbit and squash bee love story. They believe that the hoary squash bee originally used the wild buffalo gourd, *Cucurbita foetidissima*, as its main source of pollen. This bee and this gourd lived in the deserts of Mexico and the southwestern United States, thousands of years before humans developed agriculture. Over time, people began growing their own crops and domesticating the approximately twenty-two species of wild cucurbits that are native to the Americas. There's evidence that this began about 1,000 years ago. Thanks to the Holocene's increasing temperatures and improving human agricultural practices, cucurbit plants gradually spread beyond their native range, far up through North America to places where they had never grown before. The hoary squash bee, being a devoted pollinator, followed it.

These days, many varieties of domesticated squash, along with the hoary squash bee, can be found across much of continental North America, from Central Mexico to the province of Ontario, and from California to the eastern seaboard. It is an impressive geographical range, and one of the largest of any native bee in North America. (The wild buffalo gourd can still be found in the deserts of Mexico and the United States, a living reminder of the humble beginnings of an epic pollinator-host plant bond that has survived and blossomed over millennia.)

There is now evidence that these different habitats are putting selective pressures on the hoary squash bee's evolution. Long ago, the hoary squash bee lived where you could find wild buffalo gourd: in areas of desert and dry scrub. The gourd plants were perennials that grew in clumps and flowered after monsoon rains. As scattered as these patches were, they grew fairly consistently year after year. Contrast this with the fields of pumpkin and squash found today: they are often monocultures grown in high densities and watered during periods of insufficient rain. From the hoary squash bee's point of view, this provides a jackpot supply of food. However, crop rotation can make this bounty unpredictable. Squash bees that live in agricultural habitats today face very different conditions compared to squash bees 1,000 years ago.

In the course of researching these changes, scientists Nathaniel Pope, Margarita López-Uribe and their team note that domesticated pumpkin and squash plants "exude a mixture of volatile compounds that is simpler than and distinct from those produced by C. foetidissima." In other words, domesticated pumpkin and squash plants smell different from wild buffalo gourds. They found that the genome of hoary squash bees that live in eastern North America has experienced recent changes in the genes responsible for detecting and perceiving smell, as well as in the genes responsible for sensory perception more generally. The team hypothesizes that because pumpkin and squash plants in these regions are found almost exclusively in large-scale agricultural habitats, hoary squash bees who forage from them are adapting to this particular sensory environment, one quite different from that of their original habitat. This raises an interesting question: has the agricultural intensification of other crops placed similar evolutionary pressures on other wild pollinators?

Pope, López-Uribe and their team also discovered that hoary squash bees that spread into eastern North America in particular have experienced a substantial reduction in genetic diversity. This can make a species vulnerable when faced with stress, and there are a number of stressors that hoary squash bees could be currently facing. Because they rely on pollen from squash grown by humans, how humans treat these squash crops can impact the bee's health and potentially their survival.

Susan Willis Chan and Nigel Raine, two scientists at the University of Guelph, in Ontario, recently provided a demonstration of how humans were affecting hoary squash bees. They were interested in how pesticides, particularly neonicotinoid pesticides, might be affecting these bees. (Neonicotinoids, or "neonics" for short, are a group of chemicals used to protect crops from insect pests. Neonics are systemic pesticides, meaning the chemicals enter the plant's sap and travel to its stem, leaves, flowers, and pollen, protecting the entire plant. After being eaten by the insect, they attach to specific parts of the insect's brain, causing paralysis and death. Neonics seriously affect non-pest insects like bumble bees, they seep into soil and water where they are consumed by animals and plants, and repeated use increases their concentration in the environment. Neonics also persist in plants, water, and soil years after they have been applied.) They chose a farm in Peterborough County, Ontario, that had soil that was excellent for growing squash plants and for nest building by hoary squash bees. The team constructed twelve large enclosures that they called "hoop houses." Each hoop house was covered in shade cloth to keep hoary squash bees inside and other bees out. The shade cloth let in the sun and the rain, and the other elements, so it was not greatly different from the outside. The team planted acorn squash in all twelve hoop houses, but the houses differed in terms of the plants' exposure to neonicotinoid pesticides. In three hoop houses, they planted acorn squash seeds that had been treated with neonicotinoid pesticides (FarMore FI400, containing thiamethoxam). In three other hoop houses, they sprayed the acorn squash plants with a liquid pesticide (Coragen, containing chlorantraniliprole). In another three, they used a soil application of neonicotinoids (Admire 240 Flowable Systemic Insecticide, containing imidacloprid). To act as a control, the last three were not treated with any pesticides.

The team then traveled to another site and captured female hoary squash bees that were entering their nests with pollen on their legs. These "pollen pants" were a sign that the bees had mated and would soon be establishing underground nests where they would lay their eggs. Once the squash plants were in bloom, they released the female squash bees into the hoop houses. Chan and Raine then tracked the number of nests the female hoary squash bees established.

Over two years, the impact of neonicotinoids were clear. Nest construction fell by 85 percent. Five times more pollen was left behind on squash flowers. The squash bees produced fewer offspring. Yet, incredibly, the amount of acorn squash harvested was acceptable by commercial standards and gave no indication that the hoary squash bees were affected by the pesticide. I find this particularly terrifying. If growers apply neonicotinoid pesticides to soil where cucurbit seeds are planted, after two seasons hoary squash bees may disappear with no prior indication from crop yields that something is desperately wrong. "I think Canada has made a really good move, and they have eliminated the uses of drenches into soil for all neonics," Willis Chan told me. "So you cannot apply neonicotinoids directly to soil as a drench. However, they're still being applied as seed coatings. . . . It worries me."

Growers who treat their cucurbit crops with neonicotinoid pesticides could address a potentially dwindling hoary squash bee population by adding honey bee hives to their fields.

Far better to avoid neonicotinoids altogether. Honey bees would be unnecessary. In fact, for many crops, they are not needed. Pumpkins are a good example. In one study, observers found that over three years across twenty-four commercial pumpkin fields in Pennsylvania, no less than thirty-seven different bee species visited pumpkin flowers, with an average of seven species found at each field. Some were varieties of green or black sweat bees, but most often they were from several species of bumble and hoary squash bees. The field with the greatest diversity of bee species had a history of no-till agriculture, which would provide long-term, safe nesting habitat for ground-nesting species like hoary squash bees.

Although managed honey bees were one of the most frequent visitors to the pumpkin flowers, bumble bees and hoary squash bees both proved better at pollinating. "Squash bees and bumble bees actually overlap really well in their temporal niche," Willis Chan told me. They forage during the same time each day before other pollinators are active. "They're both early flyers, but at least in Ontario we don't seem to have as many bumble bees on squash plants as they do like in the northeast United States."

"I can tell you that squash bees out-compete honey bees any day of the week on cucurbit crops," said Willis Chan, "because they're up and foraging at dawn and the honey bees don't show up till . . . They might, some of them show up at 8 o'clock, but they don't peak in their numbers until 11 o'clock, and by that time all the resources have already been had by the squash bees."

Overall, wild bee populations were allowing pumpkin growers in Pennsylvania to meet their production goals. The wild bees helped produce sufficient numbers of pumpkins, and the pumpkins had an ideal size and weight. Renting honey bee hives did not appear to be necessary; wild bees were doing the job for free. Similar conclusions were made by researchers who studied pumpkin pollination in New York, Maryland, and Virginia. One grower mentioned that they decreased the number of managed honey bee hives in their pumpkin fields and saw no negative effects on yield. Of course, this depends on having healthy wild bee populations in surrounding areas, which in turn depends on providing healthy wild bee habitat. For hoary squash bees, this means chemical-free, undisturbed sandy soil where female bees can dig their nests and ensure successive generations of bees.

If you like pumpkins, thank a hoary squash bee. It is in no small part thanks to them that we can indulge our love for all things pumpkin—from pies to jack o' lanterns to pumpkin spice lattes.

Hoary squash bees nurtured North Americans' close relationship with pumpkins and other squash, yet through the development of agriculture, humans have nurtured hoary squash bees, helping them not just to survive but proliferate. "When we think of insects benefiting from and adapting to widespread agriculture, we tend to think of pests such as certain kinds of moths, flies, and beetles," said Margarita López-Uribe, associate professor of entomology at Pennsylvania State University. "By planting squash all over North America, humans created habitat for the squash bee, and that allowed the population to explode."

Hoary squash bees are an example of how the human manipulation of the natural environment can have unexpected outcomes. They also demonstrate the resiliency of nature. However, modern agricultural practices, such as chemical applications and tilling, could be creating new challenges for this. Climate change is also an issue because as a ground-nesting bee, floods can destroy its nests. Effects of climate change on squash plants will also directly affect squash bees; they may be able to find nectar from other species of flowers, but they rely on squash pollen to raise their young. We should also keep an eye on their disease and parasite levels, especially considering that hoary squash bees often share foraging areas with managed honey bees.

Recently, hoary squash bees collected from a number of farms in Pennsylvania showed high levels of two honey bee parasites, *Varimorpha apis* (formerly *Nosema apis*) and *Spiroplasma apis*, as well as a high diversity of trypanosome species. Whether these parasites are having any impact on hoary squash bees is unknown. Perhaps hoary squash bees are simply a host for these parasites and they experience no health effects—the presence of parasites does not necessarily mean that there is an outbreak worthy of concern.

It has been two seasons since hoary squash bees began visiting our garden. It is still such a pleasure for me to watch them and to know that they have some habitat in our backyard. It's still a mystery to me where their underground nests are, even though I've searched the sandy soil each year, hoping to find a little dirt pile with a pencil-width entrance. They can't be too far away. It amazes me that even though I live in a city, hoary squash bees live here too, and they found our little garden. They just show up, they are a free source of wonder and joy, and I don't have to rent them from anyone.

Chapter Six

"FIRST YOU HEAR them. Their buzz is like nothing else I've ever heard," Amy Toth, a professor at Iowa State University and an expert on bees and wasps, is describing a typical encounter with *Bombus dahlbomii*, southern South America's only native bumble bee. Then you see it. "It's so low and so big," she says. With queens that can be up to three centimeters long, "They're like these flying buses. I've studied a lot of different species before, and I had never really seen something that big and airborne. They're kind of clumsy." Besides their size and loud buzz, it's their appearance that grabs your attention. "They're just bright orange," Toth recalled. "They kind of pop out at you. Their hair is just so thick and dense. They kind of have this mammalian kind of appearance. You want to pet them because they look so plush." Doing research in the field and being in the presence of such an incredible bee never grows old for Toth. "Seeing them in person, I mean every single time, is just kind of a pleasure."

Also known as moscardón or the giant Patagonian bumble bee, *Bombus dahlbomii* was originally found across a vast region covering southern Argentina and south-central Chile, all the way to the southern edge of the continent. *Bombus dahlbomii* tends to make its home in temperate forests, but thirty years ago they were also seen in many rural areas and cities. José Montalva grew up in a town in central Chile, and he remembers seeing the enchanting giant bumble bee when he was six years old, visiting his grandfather's house. "It was in the tomato patch," Montalva recalled, "a huge, loud, fluffy orange thing buzzing around. I remember trying to grab it, but it kept getting away, although it looked too heavy to fly." Montalva is now an entomologist who studies *Bombus dahlbomii*. "In 2003, we would see thousands of the native bumble bees in the gardens of the university just outside of the capital, Santiago, where I worked. The flowers were covered with these big, fluffy, orange bees."

Bombus dahlbomii visits a number of different plant species in Argentina and Chile. It has an intimate relationship with one in particular, the wild lily *Alstroemeria aurea*. Also known as the Peruvian lily or "amancay," hybrids of this flower can be found in flower shops around the world. Between late January and early March, blooming amancay transform forests in the Challhuaco Valley in Argentina into a sea of vibrant yellow-orange blooms. Buzzing and seeming to float from flower to flower among the lenga beech trees, furry, orange *Bombus dahlbomii* worker bees would fill their bellies with nectar while pollinating the flowers. The forest would hum as flower and bee performed their intricate dance, nourishing each other in the web of life.

Then *Bombus dahlbomii* began to disappear. Montalva no longer saw it among the wildflowers on the coastal cliffs where he did annual bee surveys every year. "You hear how species like lions and rhinos could go extinct on the other side of the world, but I grew up with this one and I'm witnessing its disappearance," said Montalva. "It's happened so quickly." Indeed, *Bombus dahlbomii* is now listed as an endangered species on the International Union for Conservation of Nature Red List, the union's catalogue of those species most in danger.

What could have caused *Bombus dahlbomii*'s drastic decline? The story might have begun in the early 1980s, when about 300 non-native bumble bee queens were released at two sites in south-central Chile. The non-native bumble bee was *Bombus ruderatus*, a European bumble bee with a long tongue, introduced into Chile due to its reputation as an excellent pollinator of red clover. The queens set loose in Chile established colonies in the wild, and nearly thirty years later, *Bombus ruderatus* had spread more than 400 kilometers south from its original release sites. It could be seen on both sides of the Andes, and it became established in Argentina. During this period, there was a noticeable decline in *Bombus dahlbomii* populations in the northwest Patagonian region of Argentina.

The invasion of *Bombus ruderatus* across Chile and into Argentina was just the beginning, however. In 1997, tomato growers in northern and central Chile imported colonies of another European bumble bee, *Bombus terrestris*, or the buff-tailed bumble bee. These bees had been reared commercially in Belgium and Israel, and were sent to Chile for use in greenhouse pollination. International borders mean nothing to bees, and neither do laws in Argentina forbidding the importation of non-native species of bumble bees. Nine years later, in 2006, scientists discovered *Bombus terrestris* queens and workers on the Argentine side of the Andes near San Martin de los Andes, in Lanín National Park. This foreign bumble bee has since expanded its range more than 2,000 kilometers from its original introduction sites in Chile, and can be found as far as the southernmost tip of the continent in Tierra del Fuego and from the Pacific to the Atlantic coasts across the Patagonian steppe. This spread of *Bombus terrestris* across Chile and into Argentina is considered to be one of the most extensive invasions anywhere of a non-native species in recent times.

Scientific models predict that over time, *Bombus terrestris* will spread northward along the Andes into Bolivia and Perù, east into the Argentine Pampas, and then northeastward into Uruguay and southern Brazil along the Atlantic coast. This bumble bee is still being imported into Chile, where it not only continues to pollinate greenhouse crops but is increasingly used for blueberry pollination in open fields. Placing commercial colonies in open fields gives the bees the opportunity to establish themselves in the new areas, since new queens can leave the nest and start their own colonies. According to one report, as of 2016, 1.2 million commercial colonies of *Bombus terrestris* have been imported into Chile, more than 200,000 in 2015 alone.

The spread of *Bombus terrestris* across the two countries has been a conquest of sorts. Once *Bombus terrestris* started to be frequently seen in the wild, the first introduced bumble bee, *Bombus ruderatus*, began to decline. So did the native *Bombus dahlbomii*. *Bombus terrestris* seemed to completely replace them. How? Likely a number of factors were at play. Arguably, the biggest threat that invasive *Bombus terrestris* has brought to native bee populations is disease.

Scientist Cecilia Smith-Ramírez compares the situation to colonizers transmitting foreign disease to native peoples. "Just as

European settlers accidentally wiped out native populations in the Americas with a host of European disease such as measles, which they had no defences against, the European bumblebees can transmit pathogens that kill native bees."

Commercial colonies of *Bombus terrestris* may not show visible signs of illness. It is possible they develop resistance to certain diseases when exposed to them in the factory-like settings where they are raised, or if they have been selectively bred for resistance. So commercial bees might be carrying undetected diseases that can spread to native bee populations. This is believed to be the case with infected *Bombus terrestris* shipped to Chile. Researchers Marina Arbetman, Ivan Meeus, Carolina Morales, Marcelo Aizen, and Guy Smagghe looked at preserved specimens of *Bombus dahlbomii* and *Bombus ruderatus* caught before *Bombus terrestris* arrived in 2006. The team found no evidence in them of the pathogen *Apicystis bombi*. This suggests the pathogen did not originally exist in Patagonia. However, looking at specimens caught after 2006, all three species— *Bombus terrestris*, *Bombus ruderatus*, and *Bombus dahlbomii*—were infected with *Apicystis bombi*. The evidence suggested that *Apicystis bombi* had jumped from commercial *Bombus terrestris* to existing bumble bee populations in Patagonia. This was also the time period during which *Bombus dahlbomii* experienced its devastating population decline.

Patagonia was not the only region that was experiencing the effects of commercial bumble bees. As we saw in Chapters Two and Four, greenhouse operators in North America began using commercially produced *Bombus impatiens*, the common eastern bumble bee, which is native to eastern United States and Canada, for year-round pollination of tomato and sweet pepper. They were able to escape and mingle with wild bumble bees—some of which showed signs of population decline. Sometimes commercial bumble bee colonies are placed in open fields to pollinate berries, tree fruit, or other field crops with similar effects.

In 2015, a group of scientists in Mexico screened commercially reared *Bombus impatiens* from 120 different greenhouses. Bees from fifty-four of the locations (45 percent) tested positive for one or more pathogen.

Problems with commercial bumble bees weren't confined to the New World. In 2013, a group of scientists in England purchased forty-eight commercial *Bombus terrestris* colonies from three different producers. All three companies claimed that their colonies were pathogen-free. However, when the scientists screened the bees, only eleven colonies (23 percent) were. And bees from all three companies tested positive for one or more pathogen. When the scientists exposed healthy, pathogen-free bumble bees and honey bees to the pathogens taken from the commercial colonies, a number of the bees died.

The scientists also analyzed the pollen that had been shipped with the commercial colonies. Pollen, often from honey bees, is placed in commercial bumble bee colonies to be used as food for developing larvae until the colonies reach their destination and can go looking for it on their own. Out of the twenty-five pollen samples the scientists took, only one sample was free of every pathogen that they screened for. When they fed the contaminated pollen to healthy bumble bees and honey bees (by mixing it with sugar and water), the bees became sick and died. They also fed the contaminated pollen to honey bee larvae, and they did not survive either.

Whatever methods the commercial bumble bee companies were using to screen their colonies for pathogens before shipping were obviously not working. At the time, the United Kingdom was importing approximately 40,000 to 50,000 bumble bee colonies per year. If the study that the scientists conducted truly reflected the state of commercial colonies, then roughly 77 percent of all commercial colonies—over 30,000 each year—had the potential to spread deadly pathogens to bumble bee and honey bee populations.

Back in 2003, Dave Goulson, a renowned bumble bee expert who has written extensively on the plight of bumble bees and insects in general, wrote that we have barely scratched the surface when it comes to understanding the pathogens that affect bees. We know very little about how susceptible native bees are to those that are introduced by exotic or commercial bees. Their origins and natural range, the susceptibility of native bees to those diseases—and those pathogens' lethality—are all too often unknown.

As Goulson warned in a review of the effects of introduced bees to native ecosystems, "Studies of the incidence and identity of pathogen and parasite infestations of wild populations of native bees are urgently needed." He also reminds us that native, wild bees already exist that can pollinate our crops, and they are often more efficient than imported ones. "Countries could breed their own pollinators," he said, "but it is often cheaper to import."

Twenty years later, commercial colonies of *Bombus impatiens* and *Bombus terrestris* are still available for purchase. *Bombus huntii* became commercially available in western Canada and *Bombus vosnesenskii* is now sold commercially in California. Commercial producers of bumble bees do make efforts to ensure clean stock in their production facilities and provide guidelines for the use and containment of bees outside of their native range. However, commercial bumble bees are still deployed to greenhouses and in open fields for pollination. We have made progress in terms of the detection and identification of bee pathogens, but there is still a lot we don't know. For example, deformed wing virus and black queen cell virus, originally found in honey bees, have been detected in bumble bees, but in a number of these cases there are no outward signs of pathology. If these bees are "carriers" of the pathogen, are they contagious? If so, to what extent? At what level of infection do bees show illness or die? Are bees more susceptible to illness or death if exposure to a pathogen is combined with some form of stress, be it in the form of an additional pathogen or from any of the following: the shipping process (in the case of commercial bumble bees), habitat loss, climate change, and/or exposure to pesticides? Do all bumble bee species react the same way to pathogens as *Bombus terrestris* and *Bombus impatiens*, which are the most widely studied to date?

In 2023, a large group of scientists in the United States in Canada published detailed recommendations and best management practices for what could become a clean stock certification program for the commercial rearing of bumble bees. Recognizing that it is unlikely that viruses and other pathogens can be completely eliminated from the facilities where commercial bumble bees are reared, they state that reducing pathogen levels is still important.

One step would be to sterilize pollen used in the rearing process before bees are exposed to it. Their report stated that a number of bumble bee declines have been linked to pathogens carried by commercial bumble bees. Even if this connection is not warranted, a clean stock certification program will help build trust between end users, conservationists, and the commercial producers who have historically kept their rearing, production, and shipping methods a mystery.

Now let's return to *Bombus terrestris* in Patagonia. As an exotic species, it can place stress on native populations of bees in ways other than spreading disease. One way is by disrupting well-established pollinator-plant relationships. Some scientists noticed that *Bombus terrestris* has a liking for scotch broom (*Cytisus scoparius*). Since the arrival of *Bombus terrestris*, this non-native plant has spread more rapidly across Argentina's Nahuel Huapi National Park.

Bombus terrestris is also notorious for nectar robbing. Some flowers have a structure such that the pollinator has to reach way down inside the flower to reach the nectar. This is where having a long tongue comes in handy, because the pollinator can extend it down into the flower, sort of like using a long straw to reach liquid at the bottom of a cup. *Bombus terrestris* has a short tongue, and so it can't reach the nectar that's down inside tubular flowers. To get around this, the bee "cheats" by using its mandibles (pincer-like mouthparts) to nibble a hole near the base of the flower and it drinks the nectar from there.

Since the arrival of *Bombus terrestris*, nectar robbing has been a particular issue for several plant species, such as *Fuchsia magellanica*, a plant native to Argentina and Chile. This shrub produces vibrant red blooms that droop downward, and they are mainly pollinated by Argentina and Chile's native populations of hummingbirds. The hummingbirds' long beaks and tongues can slip down into the plant's long, tubular structure. *Bombus dahlbomii* also has a long tongue, and it is also a pollinator of *Fuchsia*. *Bombus terrestris* on the other hand, with its much shorter tongue, is unable to reach *Fuschia's* nectar the "traditional" way, so it chews a hole at the base of the flower.

I happened to witness *Bombus terrestris* nectar-robbing *Fuchsia* firsthand when I visited Ireland a few years ago. Beautiful red *Fuchsia* blooms hung over stone walls in the small town where I was staying, and huge, fluffy *Bombus terrestris* queens with their buff-colored rear ends buzzed from flower to flower. When I stopped to watch one, the bee landed on the downward facing petals and I heard *click-click-click* as the bee chewed a hole in the tubular-shaped section of the flower near the stem. The bee then stuck its tongue in the newly constructed hole and took a long drink of nectar. Once finished, it moved on to the next flower, but that flower already had a hole made by a previous bee. The queen I was watching simply landed, stuck its tongue into the pre-existing hole, and took another long drink. As I scanned the rest of the shrub overhanging the wall, I noticed that most flowers already had holes chewed in them, and a number of *Bombus terrestris* queens were putting them to good use. They were landing on the flowers and sticking their tongues into the pre-existing holes to gather nectar from the flowers. This is often referred to as secondary nectar robbing: A flower that was previously robbed of nectar is robbed again, with the new visitor using the same hole that was made and used by the original, primary robber.

Nectar robbing of the native *Fuchsia magellanica* by *Bombus terrestris* is now quite common. Marcelo Aizen is an Investigador Superior of CONICET (the National Research Council of Argentina) and Profesor Titular at the ecology department of the Universidad Nacional del Comahue in Bariloche. He has been studying plant-pollinator interactions for decades and has more recently been investigating the consequences of the invasion of *Bombus terrestris* in Patagonia. He and his team have noticed that pollination of *Fuchsia* has decreased during the same period of time that sightings of *Bombus terrestris* have become common. He suspects that it is likely due to the sheer numbers of the bee.

"Sometimes, in some species, nectar robbery can happen very, very fast," Aizen told me. "For instance, when you have low levels of [flower] visitation, for instance, by *Bombus terrestris*, you don't have nectar robbery. But all of a sudden, when the density of the bumble bees, of this invasive bumble bee, increase rapidly, nectar robbery can go from almost 0 percent to almost 100 percent in one, two, three days." During the season when Aizen along with masters student Nick Rosenberger and Aizen's colleague Lawrence Harder were observing *Fuchsia* flowers, they saw the density of *Bombus terrestris* increase so fast that colony growth alone couldn't explain the increase. They suspect that individual *Bombus terrestris* worker bees that had been foraging from other types of flowers were somehow "recruited" by their sisters to visit *Fuchsia* flowers. Perhaps back in their nest, they smelled the *Fuchsia* nectar from their sister worker bees and decided to seek out Fuchsia flowers themselves.

Scientists are not the only ones who have noticed nectar robbery of *Fuchsia* plants by *Bombus terrestris*. *Bombus dahlbomii* has noticed, too. It has been caught in the act of secondary nectar robbing. They don't usually do this because their tongues are long enough to reach the nectar inside the flower's tube-like structure. "But if there's already a hole," Toth explained to me, "they secondary rob and then they no longer pollinate." It is fascinating to see such behavioral flexibility in an insect—in this case, a native bumble bee learning how to "cheat." As Toth told me, at some level *Bombus dahlbomii* caught on to this foraging short cut. "Why take a long time to do something when you could just get a free ride on the side and skip the pollination?"

Interestingly, *Bombus terrestris* colonies tend to die off while *Bombus dahlbomii* colonies are still at their peak numbers. During the die-off, *Bombus dahlbomii* seems to recruit worker bees to *Fuchsia* flowers. These "recruits" are likely bees that were previously foraging from other types of plants, but decide to join their sisters in feeding from *Fuchsia* perhaps because of an attractive scent or some other communication. The bees extract nectar from *Fuchsia* by either extending their tongue into the flower's tube-like structure as they normally would, or sticking their tongue in holes that were chewed in the flower by someone else. Both *Bombus terrestris* and *Bombus dahlbomii*—an invasive bee and a native bee—show incredible behavioral flexibility when it comes to gathering food. *Bombus dahlbomii* in particular has shown three different responses to an invader: (1) business as usual where they drink nectar from *Fuchsia* flowers as they normally would; (2) take advantage of the holes the invaders chewed in the *Fuchsia* flowers; and (3) avoid *Fuchsia* flowers altogether and forage from something else. It's fortunate that *Bombus dahlbomii* had other plants in the region they could forage from, and that they were able to adapt their behavior. Pollinators in regions with a smaller variety of plants they can forage from would face tough competition if an invader like *Bombus terrestris* was introduced.

On some level, *Bombus terrestris* might be making nectar foraging "easier" for *Bombus dahlbomii* by giving them a foraging short-cut in the form of holes they chew in flowers. Nevertheless, they are disrupting a well-established plant-pollinator relationship between *Fuchsia* and the native orange-colored bumble bee. *Fuchsia* flowers that are not visited the "normal" way are not pollinated. No pollination harms the plant's reproduction and ultimate survival. And we must not forget the other pollinators in Fuchsia's ecosystem, such as the hummingbird *Sephanoides sephanoides*, also known as the green-backed firecrown. What impact, if any, is the high density of the invasive *Bombus terrestris*, with its nectar-robbing ways, having on it? That question has yet to be answered.

It is difficult to predict the ultimate impact of nectar robbing, however, because it depends on the pollinator-plant community in the region in question. For instance, maybe there are native pollinators who will crawl around or inside the flowers as usual, thereby becoming dusted with pollen, even though the flower has a gaping hole at its base. Perhaps the particular flower species that are robbed in a particular location or habitat are hardy enough that they can produce seeds despite being physically damaged. However, some scientists found that nectar-robbed flowers were not visited as often by "legitimate" non-nectar-robbing pollinators. In these cases, they skipped over the damaged flowers like we would avoid choosing bruised or otherwise damaged fruit in a grocery store. The result was the plant produced fewer seeds. So, that damage ultimately if indirectly, prevents seed production. No seeds mean no future generations of that plant.

There are additional ways that invasive *Bombus terrestris* can alter natural ecosystems, apart from nectar robbing. One is competition for nest sites. If new queen bumble bees emerge from commercial colonies and escape into the wild, they could occupy nest sites that would otherwise be used by wild bumble bees. It is unclear if invasive bumble bees are indeed creating a "housing crisis"; it likely depends on the area in question. For instance, there might be more potential nesting sites in natural meadows and forests than in areas that have been greatly altered by humans, such as cities. And bumble bees can be quite creative when it comes to nesting spots. They have nested under garden sheds, under steps leading to houses and buildings, in bird boxes, and other unusual places, as long as they are well-hidden and sheltered from the elements. The most bizarre place I've heard of was an old clay teapot in someone's backyard. So perhaps bumble bees will be as resourceful when competing for homes with invasive species. But again, it might be location-dependent and would also likely depend on the number of commercial bumble bees introduced to the area. Since bumble bee nests are generally quite difficult to find, it has been challenging for scientists to study the issue of nest competition and come to any solid conclusions.

Another way that invasive *Bombus terrestris* may disrupt ecosystems is by creating competition for food. An area might already be home to a variety of pollinators including various species of bees, wasps, butterflies, moths, beetles, flies, birds, and bats. Numerous studies have shown that when an invasive bumble bee like *Bombus terrestris* enters the scene, its foraging can overlap substantially with these existing pollinators. As a rough analogy, it might be like several buses filled with people descending on a small town with only one grocery store—everyone needs to eat, but the amount and variety of available food is limited.

That said, there is a lot of controversy in the scientific community over whether introduced bees really do create competition with native pollinators for flowers. It is extremely challenging to prove that competition is occurring, because bees are highly mobile, difficult to track, and nectar and pollen levels of flowers are hard or impossible to assess with the naked eye. Note, too, that "competition" in this context does not mean bees and pollinators physically fighting over flowers—although that would certainly be easier to see! Bees generally leave each other alone, even when sharing the same flower. I often see wasps, sweat bees, bumble bees, and honey bees all peacefully foraging simultaneously from the same patch of flowers in my garden. (Although I have on rare occasion seen a bee crash land into another bee already on a flower. They tussle for a second or two, but inevitably one of them flies away and that's the end of it.)

Some experts argue that just because an introduced bee shares territory with other pollinators does not necessarily mean they are in competition with them for food. For example, a type of flower might exist in a habitat that was not frequently visited by the native pollinators, but it is particularly attractive to the new bee species. The invasive bees and the native pollinators may not disrupt each other when foraging in this case, as they have separate preferences.

However, *Bombus terrestris* has several competitive advantages, especially compared to smaller, solitary bees. For one thing, it is not a picky eater. *Bombus terrestris* is known as a generalist, which means it selects food from a wide variety of flowers. Also, like other bumble bees, it can fly long distances to find food—up to at least four kilometers—and it forages early in the morning compared to other types of bees. With its large, hairy body, it can withstand cooler temperatures and rainy and windy weather compared to smaller, less hairy species of bees. It can forage at times when other bees take shelter.

Again, there is no *conclusive* evidence to date that there is competition between invasive and native bees for food. But that does not mean competition isn't happening. The odds certainly seem stacked against native bees, especially smaller, solitary bees, when it comes to finding food and sharing land with invasive *Bombus terrestris* or other species of commercial bumble bees.

During a conversation I had with scientist Marcelo Aizen about these invasive bumble bees in Argentina, he made the important point that the potential negative impacts that commercial, non-native species of bumble bees can have on ecosystems are not necessarily because the bees are non-native. Numbers matter. "The problem sometimes is not that an invasive species is bad because of being exotic or non-native," he told me. "Mostly you start having impacts when the density is very high."

Aizen explained that in Argentina and Chile, *Bombus terrestris* might not be as efficient on a per-flower-visit basis as native species of pollinators, such as *Bombus dahlbomii*, which has co-existed with the native plants for millennia. It can also cause damage by visiting flowers too often. When certain types of flowers are visited over and over by large numbers of *Bombus terrestris* (or any type of pollinator, really), this can damage their reproductive parts, leading to poor fruit quality and/or seed production, and possibly affecting future generations of the plant.

Aizen and his team witnessed the effects an overabundance of bees can have when studying raspberry fields in northwest Patagonia. Many of the fields had been supplemented with managed honey bee hives, a common practice to ensure adequate pollination. However, the area under study was also home to the invasive *Bombus terrestris*. There were so many *Bombus terrestris*, in fact, that their population density near the raspberry fields was higher than anything seen in their native ranges in Europe.

Given our inclination to assume that the more pollinators the better, we might assume that this would guarantee beautiful crops of luscious raspberries. Well, let's take a step back to examine a raspberry flower quickly. The female part of the flower consists of about seventy to ninety tiny pistils near the center of the flower. Surrounding these is a ring of anthers, or the male parts of the flower, which produce pollen. Bees or other pollinators move the pollen to the pistil. Once a pistil is fertilized, it can produce a drupelet: an individual juicy sphere, many of which together form a complete raspberry fruit. Big, plump, symmetrical fruit is produced if all pistils receive pollen. If pollen doesn't reach some pistils, these pistils can't produce drupelets and the raspberry ends up small and rather deformed-looking.

Honey bees and bumble bees visit raspberry flowers mainly for their nectar. However, by moving around on the flower, the bees increase the chance that all pistils receive pollen. The bees also transfer pollen between different raspberry flowers, another benefit to the plant.

What Aizen and his team saw in the Patagonian raspberry fields could be described as too much of a good thing. Managed honey bees and non-native *Bombus terrestris* were visiting the raspberry flowers so often that they were damaging them. The bees were breaking the flowers' styles (the stalk part of the tiny pistil), likely preventing pollen grains from traveling down the style and into the ovary at its base. Many of the frequently visited flowers grew fewer drupelets as a result, and the resulting raspberries were small and misshapen. Between the managed honey bees and the *Bombus terrestris* living in the area, each raspberry flower was being visited about 170 times per day, and in some cases upwards of 300 times per day. It only takes about ten visits for a raspberry flower to produce fruit of optimal quality.

Another factor they found contributing to excessive pollination was that the period when the flowers of the raspberry plants were at their peak overlapped with the time when bumble bee numbers were at their most numerous. Further, the variety of raspberries that Aizen and his team studied in Patagonia happened to flower after most nearby wild and cultivated plants had finished. Many hungry bumble bees at a time when only one type of flower is in bloom can result in those flowers' being exploited. And in many cases, *Bombus terrestris* nectar-robbed the raspberry buds before the raspberry flower could even open, damaging the flower and further affecting fruit production.

Bombus terrestris not only colonized the Patagonian fields, it also plundered its resources. Managed honey bees in the area also contributed to the plundering: they too visited the raspberry flowers excessively and significantly contributed to flower damage. Given the high density of *Bombus terrestris* in the area, managed honey bee hives were unnecessary in terms of pollination. (Aizen and his team note, however, that honey bee hives are often placed in or near raspberry fields for honey production rather than to achieve an optimal level of crop pollination.) It is likely that in the absence of the invasive *Bombus terrestris*, the few visits made by native bees together with properly managed honey bees would have provided a level of pollination that produced adequate yields of commercial-quality raspberries.

"Usually, in terms of plant-pollinator interactions, we have this paradigm that more is better," Aizen told me. "Not necessarily more bees is better. Perhaps I would say higher diversity is better. Not more bees of just one type. Just the biological control that exists in nature, this kind of balance is important. And that also takes to the idea that an exotic or non-native species is not necessarily bad because of being non-native. The bad thing is when it becomes very, very abundant. Trying to maintain those populations of alien species at low abundances should be a desirable and achievable goal."

To the human eye, the impacts of an invasive species might not be obvious right away. In the case of *Bombus terrestris*, it was not until the native *Bombus dahlbomii* began disappearing that people began to notice something was wrong. That was at least a decade after tomato growers started importing commercial *Bombus terrestris* colonies into Chile. If a native pollinator was being impacted, what about plant life and other living things? As Aizen mentioned to me, it's too soon to evaluate the evolutionary responses of the Chilean and Argentinian ecosystems to the introduction of *Bombus terrestris*. Evolution does not operate according to a human-centered timescale. It is possible that when humans begin to notice a significant ecological or evolutionary change, it might be too late.

In 2014, Montalva created "Salvemos Nuestro Abejorro" (Save Our Bumble Bee: https://salvemosnuestroabejorro.wordpress.com). It is a community science initiative that collects sightings of *Bombus dahlbomii* in Chile. When people see the bee, they can take a photo of it, upload it to the organization's social media platforms, and indicate where they saw it. This allows Montalva and other scientists to collect data on *Bombus dahlbomii*'s distribution. *Bombus dahlbomii*'s unique large size and bright orange color make it rather easy for ordinary people to identify, unlike other grassroots science projects that are hamstrung by the frequent misidentification of species. Salvemos Nuestro Abejorro has also collected sightings of the invasive *Bombus terrestris* and *Bombus ruderatus*. A team of scientists including Marina Arbetman, Carolina Morales, and Edwardo Zattara launched a similar community science initiative in 2021 to collect bumble bee sightings in Argentina: "Vi un Abejorro" (I Saw a Bumblebee: https://www.abejorros.ar/inicio).

Over six years, from 2014 to 2020, Salvemos Nuestro Abejorro collected more than 4,300 records. This data was used in a study by Montalva and his team to investigate whether climate change is impacting the distribution of *Bombus dahlbomii* across Chile. They found that climate change may affect the range of *Bombus dahlbomii* in the future, but over the short term, it's invasive bumble bee species that appear to be causing the greatest harm. However, they do fear that the combined effects of climate change and competition from invasive species might place the survival of *Bombus dahlbomii* in further peril in future.

Through Salvemos Nuestro Abejorro, Montalva learned that the native Chilean Mapuche peoples harvested and used honey from *Bombus dahlbomii* as a source of food, medicine, and sacred rituals. To the Mapuche people, killing the bee is taboo, even if it enters one's home. The bee could be the embodiment of the spirit of a dead relative, returning to search for their loved one.

Females in Mapuche culture held a special kinship with *Bombus dahlbomii*. The medicinal use of the bee's honey was mostly reserved for the *machis*, the wise women who cared for the sick. The bee was also revered as a goddess because of the powers attributed to its honey. Several places in Chile have names whose origin are thought to refer to *Bombus dahlbomii*. For example, Dullinco ("dullin" and "co" in the Mapudungun language, meaning bumble bee water); Misquihue (the place of honey); and Islotes Abejorros (a name thought to refer to the local abundance of giant bumble bees). As Montalva and his research team noted, the loss of *Bombus dahlbomii* is not only an ecological tragedy, "but is also adding to the cultural erosion of native culture in South America."

Montalva and other scientists including Aizen, Smith-Ramírez, Morales, Arbetman, and Toth, continue to study the impacts of the invasive *Bombus terrestris*. A major motivation for their work is to help legislators develop policies on species importation and the related global bee trade. If any good can come out of the disappearing *Bombus dahlbomii*, it is that it provides stark evidence of how the decisions made in one country about what species can be imported can have significant consequences for its neighbors. As these and other experts have stated repeatedly, we desperately need coordinated international measures to prevent further species invasions.

Bombus terrestris has conquered more than Argentina and Chile. It was introduced to, and has subsequently become established in, New Zealand, Japan, and Tasmania. The spread of *Bombus terrestris* has been one of the fastest and widest biological invasions worldwide, and compared to other introduced bumble bee species, *Bombus terrestris* is the most successful global invader. What is so special about this type of bee? What is behind its survival superpowers?

Several factors may be at play behind the success of *Bombus terrestris*. One that has been mentioned previously is pathogen spill-over. Commercially produced *Bombus terrestris* may have immunity against pathogens found in factory-like breeding conditions that wild populations of other bumble bee species do not. This may be because *Bombus terrestris* has a more "robust" immune system compared to some other species of bumble bees, or perhaps over time, *Bombus terrestris* has been able to develop immunity to specific pathogens that are present in commercial settings.

Bombus terrestris has another advantage in that its queens emerge earlier in the spring than those of many other bumble bee species, and it has a longer seasonal colony cycle in general, with colonies becoming established in early spring and not dying off until fall. In the Patagonia region of South America, its cycle is longer than those of *Bombus ruderatus* and *Bombus dahlbomii*, the two species of bumble bees it has displaced. Its early emergence allows *Bombus terrestris* to take advantage of flowers blooming at that time and grab optimal nesting sites before other species. By the time other bumble bee species wake up from their winter slumber, *Bombus terrestris* colonies may have had a strong enough head start that the other bees cannot catch up and thus struggle to survive the season.

There's another factor. Sometimes *Bombus terrestris* will produce two reproductive cycles in a year instead of one. On top of that, commercial *Bombus terrestris* colonies tend to produce more new queens each cycle compared to natural communities of the same bee. This means ecosystems where commercial *Bombus terrestris* is introduced could be inundated with *Bombus terrestris* queens, which can ultimately result in an unnaturally high number of worker bees in the region.

As mentioned previously, *Bombus terrestris* is not a picky eater and it can feed from a wide variety of nectar-producing flowers, on both native and non-native plants. Although *Bombus terrestris* has a short tongue, it works around this disadvantage by "robbing" flowers with long nectar tubes. Combine this with the high number of worker bees that *Bombus terrestris* tends to produce, and you have a formidable competitor for resources.

Another way to look at the success of *Bombus terrestris* is to flip the question around: Why are some bumble bee species declining? Certain traits might make some species more vulnerable. For instance, unlike *Bombus terrestris*, declining species tend to be pickier eaters. They tend to visit a narrower range of plants and have a less flexible diet. Habitat loss for these bees is therefore an enormous threat. If the few types of plants they forage from disappear and they are unable to switch their diets to other plants, they basically starve. Research has indeed shown that after land had been altered, certain declining species were unable to shift their foraging preferences, whereas stable species of bumble bees could.

Declining bumble bee species might also be more sensitive to warming temperatures due to climate change. Some scientists have found that certain bumble bee species more adapted for cooler temperatures, such as those found in temperate regions of the world, are gradually moving north to escape warming temperatures. This narrows the geographical areas in which they can live. Narrow ranges have also been linked to species decline, as we saw with Franklin's bumble bee.

Another possible factor making some species more vulnerable than others is their susceptibility to pathogens and parasites. For a variety of reasons, when suddenly faced with a new pathogen, the immune systems of certain wild bumble bees might not be able to cope.

Finally, it would not be surprising that declining species are victims of the synergistic effects of more than one factor. For instance, if a species of bumble bee that naturally has a limited diet is faced with both habitat loss and an influx of competing commercial bumble bees, they might be dealing with a triple whammy of less food, exposure to disease, and fewer nesting site options. Add climate change on top of this, with extreme fluctuating temperatures, droughts, and floods (and those are just the obvious effects we can see with the naked eye) and certain species of bumble bees might be tangled in a web of significant stressors, ones that stretch them to their physical and immunological limits, and they can't cope.

Is this also the case for solitary species of bees? This is an open question, as we know so much less about them.

Chapter Seven

THE TERM "INVASIVE" is usually given to species that are not native to an area, are difficult to control, and have recognized negative impacts. Given those definitions, it might seem strange to view bees—whether bumble bees, honey bees, or others— as invasive species. "Invasive" has a vaguely pejorative feel, and bees are generally viewed as beneficial.

But in some regions, as in some cities that we'll look at later, they have been introduced in such high numbers that if they were any other animal, especially a larger, more noticeable one, they would certainly be considered invasive.

The previous chapter mentioned that competition between introduced and native bee species is a complicated topic. There have been arguments over how competition should be defined and what it consists of. If there is enough food to go around—that is, enough flowers providing pollen and nectar throughout every bee species' entire life cycle—then arguably there is no competition. I've heard urban beekeeping companies advocate for more flowers to be planted in cities for this reason: if there are lots of flowers, then in theory, beehives installed in cities will not interfere with the existing pollinators that live there.

Right now, when urban beekeeping is accelerating, I think it is ridiculous to wait around until we can all agree on what competition is between bees and how to measure it. The risk posed by increasing the numbers of managed honey bees in cities is alarming to me. We saw what colonies of *Bombus terrestris* have done in Chile and Argentina. This is potentially far worse. Their colonies might hold upwards of 500 bees; one colony of managed honey bees can consist of 50,000 to 80,000. Like bumble bees, honey bees often forage early in the morning before smaller, solitary bee species appear. But unlike bumble bees, honey bees have been known to fly upwards of ten kilometers or more from their hive in search for food, farther than bumble bees. Like *Bombus terrestris*, managed honey bees are not picky eaters but are generalist foragers, gathering pollen and nectar from a variety of different flowers on their foraging trips. All of these factors combine to create a formidable threat to any existing wild bee species.

Perhaps we need to put aside worrying about "invasive" for a moment and look at the danger of "abundance" instead.

Let's have a look at southern Spain as an example. Orange groves in the provinces of Huelva and Sevilla are referred to as mass-flowering crops—they produce an enormous number of flowers all at once that last for just a brief period. To ensure pollination, orange growers place hives of managed honey bees in the orange groves. Spain is one of many countries whose managed honey bee stocks have risen steeply over the years, more than tripling since the 1960s. So, where there are orange groves in Spain, you can bet there are a lot of honey bees. So even though they are native to Spain and other parts of Europe, the significant increase in their numbers through managed beekeeping can create impacts similar to that caused by invasive species.

Adjacent to orange groves in southern Spain you can often find natural woodlands. These woodlands are home to diverse under-stories of flowering plant species, notably *Cistus crispus* (curled leaf rock rose) and *Cistus salvifolius* (sage leaf rock rose). The sage leaf rock rose is a white, five-petalled flower that blooms at the same time as the Spanish orange groves. The curled leaf rock rose is also a five-petalled flower, but it is pink or purple and blooms immediately after the orange groves.

Managed honey bees have a reputation for feeding from the most abundant floral resource that is available, and this is what scientists observed. Honey bees monopolized the two most abundant flowering plants in the woodlands, which were the curled leaf rock rose and the sage leaf rock rose, even while orange blooms were available. When the orange blooms were gone, honey bee numbers in the woodlands doubled. As a result, the native pollinators had to find food elsewhere, potentially from scarcer flowers, and ones that possibly offered pollen and nectar that differed in nutritional content from the flowers they usually foraged from.

After being visited by honey bees, both the sage leaf rock rose and curled leaf rock rose produced significantly fewer seeds compared to what they produce after being pollinated by native insects such as hoverflies, bumble bees, and solitary bees, in particular the solitary bee *Flavipanurgus venustus* (we'll call it *F. venustus* for short). Not only were the managed honey bees pushing the native pollinators from their usual food sources, but they were also impacting the survival of two major native woodland plants.

The two native flower species received large numbers of honey bee visits, but their lower seed production implies that the honey bees were not as good pollinators as the native bees. We saw this earlier with raspberries in Argentina: more visits to a flower does not necessarily mean better fruit and better seed production. Some woodland flowers were visited by honey bees fifty times or more in one day; the native pollinator *F. venustus*, on the other hand, tended to visit these only five times a day or less. How the bees visited the flower was probably a major factor as well. Besides being much larger than *F. venustus*, honey bees were more aggressive in their approach to the flowers. They tended to land directly on top of the flower's reproductive parts and collected pollen quickly, in sort of a pounce-and-grab maneuver. Instead of helping the flowers reproduce, honey bees were doing the opposite by damaging them. *F. venustus* uses a gentler strategy: it lands sideways on the outer part of the anthers, slowly walks around this, and eventually touches the stigma, transferring pollen.

Note that this was only a two-year study. We can't know now what the understory of the woodlands might look like in the more distant future if managed honey bees continue to spill over from the orange groves. Or how the local native pollinator community might change. We don't know whether the other plant species in the woodlands are being impacted by the honey bees as well. It might very well be the case that the local pollinators are not sufficient in numbers to pollinate so many orange trees, and the groves do need to be supplemented with managed honey bees. If so, we need to determine if there is an optimal honey bee hive density that can provide orange pollination while also minimizing impact on surrounding ecosystems.

This study is also an example of how crop pollination is a very narrow lens through which to view wild bee biodiversity and pollinator conservation. Approximately 300,000 flowering plant species on Earth are pollinated by bees and other animals, but fewer than 1 percent are crop plants. Without a doubt, plant-pollinator interactions—such as those between wild bees and flowers in Spanish woodlands—play a critical role in maintaining so many of the ecosystems that we take for granted.

Research conducted in the Canary Islands echoes what was witnessed in Spain. Each year, up to 2,700 managed beehives are introduced to Teide National Park in Tenerife to take advantage of the spring bloom for honey production. Scientists found that the influx of large numbers of managed honey bees on the island significantly disrupted the existing plant-pollinator networks. Wild pollinators became scarce, likely because they sought nectar and pollen elsewhere once honey bees arrived. Vertebrates that usually visited the plants practically disappeared, too, since honey bees drained the flowers of nectar. Plants that were frequently visited by managed honey bees produced fewer and/or smaller seeds or fruit, indicating that honey bees were reducing the plants' reproductive success. The scientists warned in their report that "high-density beekeeping in natural areas appears to have lasting, more serious negative impacts on biodiversity than was previously assumed."

These studies suggest that we must prioritize a deeper understanding of the complexities of bee introductions and more carefully coordinate hive placements. Ideally, we should support the wild pollinators and not install beehives at all. This is particularly the case with islands and protected habitats, where the well-being of those communities of species is often fragile and sensitive to new arrivals. In regions where native pollinators are rare locally, the argument might be made that introducing managed honey bees can help preserve the native plants that lack pollinators. However, there is the risk that honey bees will pollinate and help establish non-native plants instead. Honey bees are not like other livestock that can be forced to eat a particular food, such as grass, in a particular field. With their ability to fly long distances, honey bees can choose what and where they eat. Their flower choices may not entirely follow human plans or intentions.

Let's close this chapter with one last look at where the book began, in Spain. As we know, this is one part of the world where honey bees have been increasing over time. Carlos Herrera, an Emeritus Professor of Research at the Spanish National Research Council, has many years of field experience observing wild bees and flowers in the Iberian Peninsula. In the early part of this century he was struck by the impression from the popular media that honey bees were facing their impending demise, because honey bees appear quite plentiful where he carries out his research. In fact, it seemed that honey bees were replacing wild bees at flowers. This discrepancy between what he was reading in the media versus what he was seeing in the field nagged at him. He decided to dig into the data.

Herrera examined field studies of wild bees and honey bees involving flowers of cultivated and wild-growing plants that were performed between 1963 and 2017 across thirteen different countries on the European, Asian, and African shores of the Mediterranean Sea. Using data from the United Nations' global database, he also looked at honey bee abundance in this region over the same period. He found that over the last fifty years in the region, managed honey bee colonies increased exponentially. At the beginning of the fifty-year period, wild bees were on average four times more abundant than honey bees. At the end of it, the proportions of wild bees and honey bees had become roughly similar. Honey bees, he concluded, have been gradually replacing wild bees.

Herrera wrote in his report that the notion of a "pollination crisis" that had been blasted worldwide was inspired by the decline of managed honey bees in specific regions, notably North America and parts of Europe. Generalizing the pollination crisis and honey bee conservation actions to other areas of the world is misleading, given that in the Mediterranean Basin, managed honey bees have been *increasing*. If there is a pollination crisis there, it is the loss of pollinator biodiversity coinciding with the rise of managed honey bees. The Mediterranean Basin is a world biodiversity hotspot for wild bees and the plants they pollinate. It is home to approximately 3,300 wild bee species alone, and this number, Herrera points out, is likely an underestimate. Further, he refers to numerous studies showing that wild bees are generally better pollinators than managed honey bees—a theme we have seen elsewhere in this book. Hastily transferring conservation strategies across the world could, in the long run, unintentionally threaten biodiversity—the exact opposite of what these strategies ultimately aim to do.

More research is needed to determine what consequences this increasing abundance of managed honey bees is having on Mediterranean ecosystems. In the meantime, Herrera writes, "It does not seem implausible to suggest that, because of its colossal magnitude and spatial extent, the exponential flood of honey bee colonies that is silently taking over the Mediterranean Basin can pose serious threats to two hallmarks of the Mediterranean biome, namely the extraordinary diversities of wild bees and wild-bee pollinated plants."

Herrera's work, along with what we have seen in previous chapters, shows how the "pollination crisis" and calls to "save the bees" can mislead by misrepresenting a situation that is extremely nuanced and complex. What problems bees face depends on what region of the world you are looking at and what species of bee you are talking about. As we have seen, pollination is so much more than honey bees. Honey bees, and other managed bees like commercial bumble bees, are possible vectors of disease and other threats to natural pollination networks. Yes, in some cases, we need managed bees to help pollinate the massive amounts of food we require, but depending on the region, wild bees might actually be able to do the job for free. But this all depends on where you're looking. The loss of managed honey bee colonies in some regions can be due to a variety of factors, perhaps with two or more factors acting synergistically. A simple solution is unlikely when considering such complexity and nuance.

And yet, intentionally or not, today a simple and very misguided solution is being promoted around the world.

Chapter Eight

ONE SATURDAY MORNING I attended a workshop at my local library branch. I had signed up for their waiting list and there was an opening at the last minute. When I entered the room, all seats were taken, and the librarians had to add more. I stood in a back corner with a few other people and surveyed the crowd. A diversity of people were in attendance. We were arranged around the edge of the room, because the center was taken up by a large table covered with equipment and what looked like an enormous, stainless steel cooking pot. Leaning up against the table were large posters printed on rigid board. "Meet the members of the hive!" exclaimed one. "The Widespread World of Bees," said another. The last one on the end was titled "Backyard Bees of North America." The table and posters were all sitting on plastic sheeting that had been spread over the floor. The setup implied we were about to do something messy.

The workshop was called "From Hive to Honey Jar." Two representatives from a large turnkey urban beekeeping company were there to show us how to get jars of honey from a beehive. First, they passed around a smoker—a metal, hand-held tool that looked a bit like a small kettle. It was filled with smouldering burlap, and the reps explained that beekeepers use smokers to calm honey bees when they open the hive. They said the smell of the smoke masks the alarm pheromone that some of the honey bees emit when they discover an intruder.

As the smoker made its way around the room, they next passed around a wooden frame from a beehive filled with wax-capped cells. Although the company had recently installed a beehive on the library's property, this frame was not from there. The library's frames were being processed elsewhere, and this was from some other hive. While each person in turn peered at the frame of waxy cells, one of the reps explained the lifecycle of the honey bee, the social structure of the colony, and other honey bee-related facts.

Next came the chance for more audience participation. The frame of wax-capped cells was propped up on the large central table. Each person was invited to step up and use a metal hand-held tool that looked somewhat like a hair pick, to gently and slowly scrape the layer of wax off the cells. This revealed the oozing, golden honey underneath. One of the reps stood close by to demonstrate how to scrape the cells for those who were unsure, and she also collected the scraped-off wax and placed it in a bucket.

When my turn came, I was greeted with the heady, thick scent of honey. As I gently pushed the now-sticky tool from the bottom to the top of the frame, revealing more juicy, golden honey, I thought about how incredible it was that each of the hundreds of little cells I was now exposing had once been carefully tended by honey bees. They had filled each individual cell and then meticulously sealed it for safe keeping. The honey was different colors, too. The honey in the top section of the frame was bright yellow, whereas the honey near the bottom was a more earthy-brown. It made me think of peeling back the lid of a new tub of Neapolitan ice cream, and seeing the distinct stripes of chocolate, vanilla, and strawberry.

Once everyone had a turn at scraping off the wax, it was time to get the honey out of the frame. Enter the enormous, stainless steel cooking pot contraption. The reps explained it was a hand-held spinner. They inserted the frame inside it and demonstrated how to turn the handle, like a salad spinner or centrifuge. The frame spun around and around, flinging the honey out of each of the little wax cells and onto the inside walls of the pot-like container, where it then dripped and oozed to the bottom.

Everyone crowded around the spinner, watching with excitement as each person had a chance to crank the handle. Those with more arm strength turned the handle faster, the frame whipping around and around, strings of honey flying off like sticky, gooey sparks. The plastic sheeting on the floor did its job, catching spatters of honey that escaped the spinner. People had their phones ready and were taking photos and videos, announcing what great posts they would make on social media.

The grand finale came when the reps gave each person an empty five-ounce jar. One at a time, we stepped up to the spinner to have our jar filled with the honey we had just spun out of the frame. There was a palpable collective reaction of amazement and wonder as we watched the golden liquid slowly pour into the jars.

"Isn't it beautiful?"

"What a color!"

"So exciting!"

"It's like tapping a honey keg."

One of the reps, standing near the "honey keg," held a roll of labels and stuck one on each person's honey-filled jar. The label advertised the turnkey urban beekeeping company's name. Like the rest of the crowd, I couldn't help but gaze and marvel at the liquid gold inside the tiny jars, like we had all discovered a hidden treasure.

What is a turnkey urban beekeeping company and why was it providing a workshop at my local library? Turnkey urban beekeeping companies rent out managed honey bee hives to individuals, organizations, and schools in urban settings. They install one or more hives on the client's property and the company's team of trained urban beekeepers look after all of the maintenance and management of the hives. At the end of each season, they provide the clients with the honey that their hive or hives have produced. The larger companies offer a variety of other incentives: workshops, team-building activities, real-time updates on the hive's/hives' health through an online platform, corporate gifts to be used as marketing tools, and landscaping consultations. Some also offer honey and pollen analysis to identify the species of flowers the honey bees had foraged from, and to check that their honey bees are healthy. Some of the companies refer to these analyses as "biomonitoring."

Two large turnkey urban beekeeping companies I spoke with had very humble beginnings. One was founded by three friends who once performed as a jazz trio. Back in 2010, one of their uncles invited them to spend a summer at his farm, where he kept honey bee hives. After working with the honey bees that summer, the trio was hooked. They set up their own hives when they returned to the city. Three years later, they launched their turnkey urban beekeeping service. By 2021, they had installed nearly 3,400 beehives in major cities across North America and Europe. Many of their clients are corporations. Others are schools and public institutions, like the library near where I live. Representatives from the beekeeping company told me that in the cities where they have been present the longest, like Toronto and Montreal, they have about 400 hives in total.

"We're an international urban beekeeping company focused on education," one of the representatives told me. "There's one hive per site, and it's managed more for education and engagement and biomonitoring than it is for producing tons of honey." As one of the company's founders has said, "We wanted to try something new while reconnecting people to nature."

Although reconnecting to nature is still a main focus, the company also emphasizes creating greener cities and helping corporations engage with their community while meeting their sustainability goals. According to their website, they do this through offering to track the impact of their clients' buildings with "science-based biodiversity data." This is where the "biomonitoring" comes in. Several times a year, honey and pollen from the client's hive(s) are sampled and analyzed. From these samples, the company determines the diversity of flowers in their customer's area, pollen quality, and indicators that their honey bees are healthy. From this, they provide the client with an "environmental report." The client also receives recommendations to improve "biodiversity" around their property. Finally, the client's data is confidentially shared with a global network of scientists to "support research on pollinator health and protection."

Honey and pollen analysis along with landscaping recommendations are ways that the company can continue to engage clients after they have had a beehive on their property for a while. "The way that our clients grow from a sales point of view," one representative told me, "is not in adding more hives, it's in more like improving their landscaping practices or distributing like native wildflower seeds. And we go to great lengths to make sure that each package of seeds that we're distributing are native to each different region."

Another turnkey urban beekeeping company began with a graduate student who was studying honey bee immunology around the time that Colony Collapse Disorder was hitting the news. "Then the economy tanked," he told me, "and I was getting ready to graduate. And I thought, ugh, this is a bad time to graduate with a very specific degree. I was bar tending. I started teaching adjunct classes. And I had nothing to lose. I started a Facebook page. And I said, 'I'm selling beehives. Does anybody want one? I'll volunteer my time to manage them in exchange for research funding. You keep the honey, we get the data.'"

He began renting out honey bee hives from his apartment in 2010. Fast forward to today, and his company now installs and maintains thousands of honey bee hives in major cities across the United States. He has residential and corporate clients, and he has a network of over 100 beekeepers. He assured me that his passion for honey bee health is still at the heart of what he does. He perceives each beehive as a data-producing node in a network used for scientific research to understand and improve honey bee health. "We've collected a data library using a standardized beekeeping method that has about 85,000 beehive visit records," he explained to me. "And what's unique about that approach is with the science, we wanted beekeepers to do the same thing so that we could understand if the variation that we're seeing in bee health metrics was due to the bees alone and not influenced by beekeeper method."

Like the previous company, this one also analyzes honey samples to determine what flowers the honey bees are visiting. They then use this information to recommend to clients what to plant. "Once we look at the plant lists from the honey bee hives that are giving us research data, then we're creating those habitats to create more flowers and forage," he told me. They offer their clients a variety of services and incentives including year-end reports on honey and bee health, impact and sustainability reports, honey tasting events, and of course, honey and custom honey labels.

Across North America and Europe, there are a number of urban beekeeping operations much smaller than the two just described, and their intentions may differ as well. Depending on who you talk to, urban beekeeping can be a controversial topic. One area of controversy is whether urban beekeepers are—knowingly or not—benefiting from the "save the bees" narrative that gathered steam after media coverage of Colony Collapse Disorder.

I spoke with a beekeeper who described himself as a "sideliner." He has less than 100 hives and uses his honey bees as an additional source of income. He is suspicious of some of the urban beekeeping companies. "You see it all the time," he told me. "This 'save the bees' narrative. Honey bees are not endangered. They are a managed species. Stop making money, either tugging on people's heartstrings that don't know any better . . . Beekeepers know better, though."

"I know friends of mine who make a handsome living," he continued, "renting out—they call them host hives—and people, even just normal people are paying thousands of dollars to have a couple hives in their backyard," he told me. "I recently read the term *bee washing*, and to me it sounds . . . It's very greasy. I wish maybe I could bend my values a bit and make some of that money. But yeah, I think it's dishonest and greasy."

Bee washing is a term coined by scientists J. Scott MacIvor and Laurence Packer to describe products or actions that claim to support declining bee populations without due diligence or scientific support. It is a spin on the term *greenwashing*, which refers to claims that something is more environmentally friendly, or less environmentally harmful, than it actually is.

MacIvor and Packer came up with the term bee washing after investigating the effectiveness of bee hotels: structures that have been mass-marketed in North America and Europe to support pollination and wild pollinator conservation. Also called trap-nests or nest boxes, they are modeled after the above-ground nests that some solitary bee species prefer—hollow, pithy stems and beetle burrows in wood. Bee hotels can be made from bundled plant stems, paper tubes, or holes drilled in wood or molded plastic.

Despite their growing popularity, MacIvor and Packer, both bee experts, were unaware of any evidence to support the idea that bee hotels work. So, they decided to check them out themselves. Over three years, from 2011 to 2013, they placed 600 bee hotels across the city of Toronto, and monitored what critters nested in them. It turned out the most frequent tenants were non-native wasps. These wasps occupied almost 75 percent of all bee hotels each year.

MacIvor and Packer found that wasps might not be the only unintended guests that bee hotels attract. They could be hotspots for disease. Cavity-nesting bees of either the same or different species don't usually nest close together naturally, but bee hotels encourage it. This allows parasites and/or disease the chance to spread both within and between species. If bee hotels have thin-walled nest tubes, this could allow for parasites to travel within the hotel. Encouraging cavity-nesting bees to nest together also creates a target for predators. Bee hotels have the potential to become death traps. They might also just simply not attract bees.

That connects with the fact that we don't know much about the best possible design for a bee hotel. The ideal diameter and length of nesting tubes to attract native bees to the proper moisture levels and whether it's even possible to mimic the same moisture balance that occurs in natural cavities. Too much moisture in artificially constructed nesting sites could lead to bees dying from mold.

MacIvor and Packer concluded that according to their findings, bee hotels are an instance of bee washing. Retailers and promoters have claimed, irresponsibly, that bee hotels are a way to help native bees when instead, MacIvor and Packer found bee hotels mostly supported non-native wasps. At their worst, bee hotels may act as population sinks for bees, accelerating their demise through parasites, predation, mold, fungus, and disease. MacIvor and Packer suggested that bee hotels could potentially be designed to be useful tools for conservation biologists and conservation-minded citizens, but more research is necessary. It is also worth noting that although retailers and bee hotel advocates target cavity-nesting bees, the majority of bee species around the world nest in the ground.

Unfortunately, bee washing is not limited to bee hotels. Karin Alton and Francis L. W. Ratnieks, two scientists at the University of Sussex, in the United Kingdom, found that a number of products sold to help bees and other pollinators, such as bee bricks, insect "homes," and seed balls, were more gimmick than effective wildlife helper. They recognized that the cost of proper scientific testing is often outside the scope of small businesses, but still, many products were misleading and simply did not have the evidence to support their claims. Alton and Ratnieks also cautioned *caveat emptor* for "award-winning" products. In many cases, products' awards were for sales and marketing achievements rather than for actually helping pollinators. "It would be convenient if supporting wildlife was as simple as buying an item and putting it in the garden," Alton and Ratnieks wrote in their report. "However, the results of our investigation indicate that many of the products on sale may not be much help at all." We have seen how complex and nuanced bee health and bee declines are. As Alton and Ratnieks note, it is highly doubtful that products alone can help them.

The promotion of urban beekeeping is also arguably a form of bee washing. For instance, large turnkey urban beekeeping companies often use the terms "sustainable" or "sustainability" in their advertising, but research shows that in a number of cities, the upward trend in urban beekeeping is *not* sustainable. Scientists Joan Casanelles-Abella and Marco Moretti compared the number of beehives across fourteen Swiss cities between 2012 and 2018, and found that during that time these increased from 3,139 to 6,370.

The average density was about eight hives per square kilometer. Considering each additional hive contains roughly 50,000 honey bees, that's a heck of a lot of honey bees.

Casanelles-Abella and Moretti then used modeling to assess whether the cities have enough urban greenspace (i.e., flowers) to support the increasing numbers of managed honey bees. The answer was no. According to their calculations, in most of the fourteen cities, there was not enough urban greenspace in 2012 to support the original number of hives. The research team looked at only managed honey bees; they did not even consider other pollinator species that rely on the cities' flowers. Other researchers found a similar situation for London, United Kingdom: The city simply does not have enough floral resources to accommodate the recent boom in urban beekeeping. "Our work was inspired by work done by the London Beekeepers Association," Casanelles-Abella told me. "They were the ones that sparked our research by asking if there were too many hives in central London."

The large turnkey urban beekeeping companies described previously do claim to offer landscaping recommendations to their clients. Would this help offset any potential shortage of flowers caused by the influx of managed honey bees? Their honey and pollen analyses would indicate only what species of plants the *honey bees* are visiting; they do not necessarily show what plants are needed by existing wild pollinators like bumble bees and solitary bees. One university professor told me her team has done calculations on how many plants would be needed for each hive these companies install and, given the rate at which honey bee hive densities are increasing in cities, it would be in the millions. Interestingly, she has had trouble getting this research published.

"Greener cities" and "positive environmental change" are other buzzwords used by turnkey urban beekeeping companies. I have yet to see any peer-reviewed, scientific research confirming that urban beekeeping results in greener, more environmentally friendly cities. We have seen how managed honey bees can disrupt natural plant-pollinator networks, how they have the potential to be vectors of disease, and how they are not the best pollinators for a number of plant species. Increasing the density of managed honey bee hives—in some cities, by a substantial amount—has the potential for major negative environmental impacts.

Scientists are discovering that cities can host a surprising diversity of wild pollinators. Thanks to bylaws preventing pesticide use, a mix of possible nesting locations in gardens, parks, and yards, and the variety of plants people grow in their planters and gardens, cities can be a refuge of sorts for wild bees. Unless they are flooded by managed honey bees. A clear example of this happened in Montréal, Canada's second-largest city. Researchers had carried out a large bee diversity survey ending in 2013. In the years following, there was an enormous increase in urban beekeeping, with almost 3,000 managed honey bee hives installed in the city. Researchers Gail MacInnis, Etienne Normandin, and Carly D. Ziter from Concordia University and the University of Montréal surveyed the wild bee population again in 2020, after the influx of honey bees. They found that as honey bee abundance increased, wild bee diversity decreased significantly. This was particularly the case for smaller species of bees. MacInnis, Normandin, and Ziter suggest that smaller bees might be more vulnerable as they cannot fly as far to collect food, and therefore cannot expand their foraging range when faced with an invading abundance of honey bees.

There have been similar findings elsewhere. The number of managed honey bee hives in Paris, France, more than doubled between 2013 and 2015, from around 300 to 687 hives. This equates to about 6.5 hives per square kilometer. Similar to what was found in Montréal, this surge in urban beekeeping was accompanied by a significant decrease in wild pollinator activity. This was especially seen with large solitary bee species, bumble bees, and beetles. If urban beekeeping companies claim to promote greener, more environment-friendly cities, emerging research suggests this is not the case when it comes to supporting wild pollinators.

Sheila Colla, a scientist we have met several times in this book, has written about the consequences bee washing can have on wild bee health and conservation. Too often it features overly simplified information, or even intentional misinformation, that can harm, rather than help, bees. In my view, a perfect example is how urban beekeeping companies have used the term "biodiversity." Claiming that honey analysis provides "biodiversity data," and that their landscaping recommendations will "improve the quality of biodiversity around your property" provides an overly simplistic and misguided view of what biodiversity is. Honey analysis shows what plants *honey bees* from *that particular hive* visited. Period. It says nothing about other species of plants and pollinators, and their already-existing relationships. Biodiversity encompasses the entire variety of life you can find in an area, from insects and arachnids to birds to mammals to reptiles to amphibians to fish to plants to trees to fungi and even microorganisms like bacteria. Every living thing, and all the interactions among them. It extends far beyond honey bees and the flowers they like.

Connecting managed honey bees to ideas about local biodiversity is also extremely misleading because managed honey bees are livestock. Humans manage them for their own use. We selectively breed them (for example, to be more docile and less likely to sting, or to exhibit hygienic behaviors), and they are intended to thrive in human-made environments: agricultural landscapes in the case of commercial beekeeping, or cities in the case of urban beekeeping. As discussed earlier, humans keep honey bees for pollination, honey, wax, and companionship. Thomas Seeley, a world-renowned expert who has researched and written extensively about the behavior and social life of honey bees, states in his book *The Lives of Bees,* "Honey bees share with dairy cows the fate of being economically important animals that are thoroughly manipulated by humans to boost their productivity." In North America, honey bees are a non-native, introduced species. Urban beekeeping companies may claim they are "connecting people to nature," but they are in fact connecting people to livestock.

Further, as Casanelles-Abella and Moretti point out, beekeeping is a special case of livestock raising. Unlike cows and pigs, where humans provide their food, honey bees can move around freely and choose food for themselves. As we saw earlier, even bumble bees kept in greenhouses cannot be contained. Honey bees also reproduce faster than other livestock.

Unlike livestock that graze, trample, and otherwise alter the land in obvious ways, beekeeping may not seem like it is exploiting the environment. Casanelles-Abella told me that we can't look out into a field and see that the flowers have been drained of nectar. We can't look out and see which flowers have been pollinated.

Early on in this book we said that, thanks to humans' long connection to the honey bee and their obvious benefits—pollination and most notably honey—the assumption is the more honey bees the better. But what if there aren't enough plants to feed the bees? What if the honey bees end up pollinating the "wrong" plants, ones that, like them, are invasive species?

Urban beekeeping does not support biodiversity because the focus is on one species and on increasing the numbers of one species. To my knowledge, turnkey urban beekeeping companies do not monitor wild bee populations in the areas where they install their hives. Installing more and more managed honey bees in cities has the potential to replace the existing diversity of wild bees both there and in surrounding areas with only one type of bee. Biodiversity—a rich variety of life in a region—provides ecosystems with resiliency. This is particularly important these days with the challenges presented by climate change and pathogen spillover from managed species. Biodiversity provides insurance against events like disease, storms, and wildfires. Increasing the number of managed honey bee hives in an area is the opposite of supporting biodiversity. It creates a more homogenous and vulnerable ecosystem.

Equating managed honey bees with biodiversity also implies that when it comes to promoting and protecting biodiversity, installing managed honey bee hives is "enough" and so is providing the plants they are attracted to. Oversimplifying and misrepresenting the concept of biodiversity runs the risk of fostering complacency and minimizing the need for evidence-based decision-making.

Turnkey urban beekeeping companies may claim that bee health or pollinator health is one of their priorities. Arguably, this is another case of bee washing. "If they are serious about bee health," Colla told me, "they would be transparent about disease levels their hives have and not skirt around pathogen spillover as one of the main threats, if not *the* main threat, to wild bees."

Urban beekeeping companies portray themselves as bee experts and gain trust from their clients as such. If the companies are aware of the seriousness of pathogen spillover, how are they addressing it? If they don't talk about pathogen spillover or disease levels of their hives with the public, then the public may not see pathogen spillover and disease as a threat. Like biodiversity, the concept of "bee health" has been oversimplified and how it's talked about is misleading. Colla cites the precautionary principle: "When it is scientifically plausible that human activities may lead to morally unacceptable harm, actions should be taken to avoid or mitigate that harm: uncertainty should not be an excuse to delay action."

No matter how well-intentioned turnkey urban beekeeping companies might be, or how humble their beginnings, these businesses are bee-washing the public. It is the scale at which the bigger companies operate that is particularly alarming. Each one of their thousands of hives installed across North American and European cities adds roughly 50,000 or more honey bees to the environment. Would we allow the same to happen if it were any other animal?

In 2022, bee scientist Zach Portman wrote a short online essay titled "How I'm Helping to Save the Birds by Keeping Chickens." In it he says that, worried about declining wild bird numbers, he has hit on a solution: he'll keep chickens. This obviously "bird-brained" idea is based on what he calls "the misguided practice of keeping honey bees to save the bees." Portman's essay is also meant to encourage us to think more deeply about the actions we take to help the environment, our power over other species in general, and the responsibility we have for them. (With his permission, I have included Portman's entire essay in an addendum at the end of this book.)

When I attended the "From Hive to Honey Jar" workshop, the excitement, attentiveness, and engagement I felt from the crowd was remarkable. Whenever there was a pause in the reps' instructions or presentation of honey bee facts, they were peppered with endless questions. Yet the whole time, the poster on the floor that was propped up against the table, illustrating the variety of wild bees you can find in your backyard, stood like a lonely footnote that everyone glossed over. The reps referred to wild bees a few times, but again, it was like a footnote. Honey bees were obviously center stage. The fact that people were getting their hands sticky and bringing something home from the honey bees just cemented the honey bee's popularity even more.

I thought about the garden in my own urban backyard—a work-in-progress each year as I sharpen my gardening skills. It's not a big garden, but I am always amazed by the variety of visitors I see. I've seen several different species of bumble bees, green metallic sweat bees, and tiny solitary bees that I have yet to identify. And then there's the squash bees that appear each early morning and by noon are cozied-up together in the closed squash blooms. Spending even just a few minutes each day watching them, discovering who dropped by to visit this modest little garden, has become a highlight of my day. In their company, I feel calm, at peace. Dare I say, to borrow a phrase from the urban beekeeping companies, I feel "connected to nature." Real nature. No honey required. No monthly fees. All it cost me were some flowers and a bit of time to plant them.

Conclusion

WHILE DOING RESEARCH for this book, I came across a quote that stuck with me. It was in an article Sam Knight wrote for *The New Yorker*, and the quote was from a beekeeper who abstains from taking honey from his beehives. Referencing the movie "When Harry Met Sally," he said, "There was this line, 'Sex always gets in the way of friendship.' I think honey always gets in the way of us appreciating bees." I would push it even farther: it's hard to appreciate your friends (bees) when you continuously steal what they make (honey) and pimp out their services (renting them to urban clients).

People can now mass-produce and distribute "good things" on an incredible scale. This includes honey bees and bumble bees, as well as products like bee hotels. Despite any good intentions, the consequences of racing ahead with mass production and distribution without evidence-based decision-making can be too grave—for ourselves and for the planet. The continuous promotion and proliferation of urban beekeeping is not justified by the argument or the mindset that bees are "just insects." On the contrary, bees are complex organisms, in terms of both body and mind. If you need convincing of this, I'll again recommend Lars Chittka's *The Mind of a Bee* and Stephen Buchmann's *What a Bee Knows*.

Wild bees desperately need habitat. This includes nesting sites as well as flowers. You don't need to be an expert gardener or have an extensive garden to make a difference. Leaving bare soil undisturbed can provide a home for ground-nesting bees. If you don't have a yard, one or more planter pots or planter boxes offering native flowers from early spring to fall can be an important source of food for wild bees living nearby. A warning when it comes to flowers: a number of plants have been bred so their flowers look nice, but they may not provide the high-quality nectar and pollen that bees need. There are many excellent resources available that can guide you in terms of what native plants bees are attracted to in your area, how to grow them, and ideas for how you can create a pollinator-friendly garden. If you live in the Ontario and Great Lakes region, I highly recommend the book *A Garden for the Rusty-Patched Bumblebee* by Lorraine Johnson and Sheila Colla, illustrated by Ann Sanderson. Also available from the same authors and illustrator is *A Northern Gardener's Guide to Native Plants and Pollinators: Creating Habitat in the Northeast, Great Lakes, and Upper Midwest*.

If gardening isn't your thing, there are a number of existing pollinator-themed citizen science and volunteer opportunities. Have a look online to see if there is anything available in your area. Also, keep in mind that steps we take to slow climate change ultimately help the bees. Whatever you decide to do, I urge you to be vigilant for clever marketing and bee washing tactics if there are future "helpful" innovations that become available for purchase.

I sincerely hope that Jode Roberts giving up urban beekeeping will be part of a growing, global trend. For the bees' sake, and for the sake of the planet.

Addendum

How I'm Helping to Save the Birds by Keeping Chickens:
It's All About Boosting Bird Numbers!
By Zach Portman

A while ago I heard that birds are declining all over the world, and I found that really upsetting. So I decided something needed to be done and I am exactly the person to do it.

I thought about the best way I could help and did a couple google searches, and ultimately I decided to get some backyard chickens. This way, I'm boosting the number of birds and doing my part to help make sure birds don't go extinct. Plus, I get the added benefit of fresh delicious eggs! It's a win for everyone: the birds are saved, I get to personally benefit, and I look good while doing it.

One of the best things about chickens is that they're so easy to get. You can just go buy them! That came in really handy when most of my chickens got some kind of disease and their feathers fell out and they died. In fact, every few years I cull off the old chickens and buy new ones to make sure my chickens stay nice and healthy and produce lots of eggs.

Some people have said to me, "aren't you personally benefiting from this because you get the eggs?" Those people couldn't be more wrong. Any benefit I get from eggs pales in comparison to the wonderful knowledge that I am helping the environment and saving the birds.

Other people have said to me, "But aren't chickens not native to this area?" They may not be technically native, but they've been here for hundreds of years so they are practically native. By pointing out that a chicken isn't "native" or "wild" you're just being divisive. Why can't you just accept that we're all working toward the same thing, which is saving the birds? We'll accomplish so much more if we work together rather than focusing on our differences.

Plus, keeping chickens has provided a gateway into the wonderful world of birds. Just the other day, I noticed a bird in my neighbor's yard! Since I've started keeping chickens, my eyes have been opened to this amazing world. When I'm out feeding my chickens and gathering eggs, and doing regular checks for pests and disease, I just feel so connected to nature.

Due to my efforts, there are now 30 more birds in the neighborhood. A recent study showed that there are three billion fewer birds in North America than there used to be. So if everyone in the United States, Canada, and Mexico kept just six chickens, that would bring the numbers back up to where they used to be!

Recently, I've even started a company where other people or corporations pay me to keep chickens on their property and give them the eggs. This way, they can help save the birds too (and get fresh local eggs!) without all the hard work. The only thing they have to do is send me monthly cash payments. They also have the added benefit of being able to post pictures of their birds on social media so that everyone can know how much they are helping the environment. This is really important because some of these companies have really horrible track records when it comes to the environment, but by keeping chickens and boosting birds, they're able to wash away any bad stuff they've done and show off what good environmental stewards they really are.

One of my neighbors claimed that because I have so many chickens in my neighborhood now, they are eating all the bugs that are needed to feed the other birds. I really don't understand that. I have never seen one of my chickens chase or steal food from another bird, so I don't see how they could be competing or having any other negative effects. Plus I've had barely any disease outbreaks among my chickens, so they probably aren't spreading anything to wild birds.

I hope you'll join me in keeping chickens to save the birds. It's super easy. You can even order them online and have them shipped through the mail! The benefits are endless—fresh local eggs, and the knowledge that you are making a difference and helping the environment. Together we can boost bird populations and prevent birds from going extinct.

References

Introduction

Jode Roberts giving up urban beekeeping:

Roberts, Jode. "About Jode." Retrieved November 25, 2023, from https://www.joderoberts.com/about.

Roberts, Jode. "Why I'm Giving Up Beekeeping: Urban Beekeeping Is, Paradoxically, Bad for Bees." *Toronto Star.* (May 23, 2023). Retrieved November 25, 2023, from https://www.thestar.com/opinion/contributors/why-i-m-giving-up-beekeeping-urban-beekeeping-is-paradoxically-bad-for-bees/article_451ec774-6c02-5b59-bbc1-1a4d068a59bd.html.

Chapters One and Two

Abivardi, Cyrus. "Honeybee Sexuality: An Historical Perspective." *Encyclopedia of Entomology.* Dordrecht: Springer, 2004. https://doi.org/10.1007/0-306-48380-7_2057.

Aristotle. *The History of Animals* (Book V, Part 21). Translated by D'Arcy Wentworth Thompson. *The Internet Classics Archive*, 2009. Retrieved April 7, 2022, from http://classics.mit.edu/Aristotle/history_anim.html.

Aristotle. *The History of Animals* (Book IX, Parts 40 and 43). Translated by D'Arcy Wentworth Thompson. Retrieved April 7, 2022, from https://penelope.uchicago.edu/aristotle/histanimals9.html.

Buchmann, Stephen. *Letters from the Hive: An Intimate History of Bees, Honey, and Humankind.* New York: Bantam Books, 2005.

Costa, James T. *Darwin's Backyard: How Small Experiments Led to a Big Theory.* New York: W. W. Norton & Company, 2017.

Crane, Eva. "The World's Beekeeping—Past and Present." In Grout, Roy A., Ed. *The Hive and the Honey Bee*, 1–10. Hamilton, IL: Dadant & Sons, 1963.

Crittenden, Alyssa N. "The Importance of Honey Consumption in Human Evolution." *Food and Foodways* 19 (2011): 257–273.

Darwin, Charles. *The Origin of Species.* Hertfordshire: Wordsworth Classics of World Literature, 1998.

Gunda, Béla. "Bee-Hunting in the Carpathian Area." *Acta Ethnographica Academiae Scientarium Hungaricae Tomus* 17 (1968): 1–62.

Lickers, Henry, March 4, 2019. Personal interview.

Maderspacher, Florian. "All the Queen's Men." *Current Biology* 17 no. 6 (2007): R191–R195. https://doi.org/10.1016/j.cub.2007.02.017.

Michez, Denis, Maryse Vanderplanck, and Michael S. Engel. "Fossil Bees and Their Plant Associates." In Patiny, Sébastien, Ed. *Evolution of Plant-Pollinator Relationships*, 103–164. Cambridge: Cambridge University Press, 2012.

Obituary: Frederic William Lambert Sladen. *The Canadian Entomologist* 23 no. 10 (1922): 240.

O'Toole, Christopher. *Bees: A Natural History.* Richmond Hill, Ontario: Firefly Books, 2013.

Poinar Jr., George O., and Bryan N. Danforth. "A Fossil Bee from Early Cretaceous Burmese Amber." *Science* 314 (2006): 614. https://doi.org/10.1126/science.1134103.

Prendergast, Kit S., Jair E. Garcia, Scarlett R. Howard, Zong-Xin Ren, Stuart J. McFarlane, and Adrian G. Dyer. "Bee Representations in Human Art and Culture through the Ages." *Art & Perception* 10, no. 1, (2021): 1–62. https://doi.org/10.1163/22134913-bja10031.

Shakespeare, William. *A Midsummer Night's Dream.* Edited by Sukanta Chaudhuri. Arden Shakespeare, 3rd ser. London: Bloomsbury, 2017.

Shakespeare, William. *Henry V.* Edited by Barbara A. Mowat and Paul Werstine. New York: Simon & Schuster, 2020.

Sladen, Frederick William Lambert. *The Humble-Bee: Its Life History and How to Domesticate It.* London: MacMillan and Co., 1912.

South African National Biodiversity Institute (SANBI). "African Honeybee." (2022). Retrieved March 27, 2022, from https://www.sanbi.org/animal-of-the-week/african-honeybee/#:~:text=The%20African%20honeybee%20subspecies%20(Apis,in%20the%20winter%20rainfall%20area.

Stamp, Jimmy. "The Secret to the Modern Beehive Is a One-Centimeter Air Gap." *Smithsonian Magazine.* (September 6, 2013). Retrieved April 4, 2022, from https://www.smithsonianmag.com/arts-culture/the-secret-to-the-modern-beehive-is-a-one-centimeter-air-gap-4427011/#:~:text=In%201851%2C%20Reverend%20Lorenzo%20Lorraine,and%20designs%20developed%20over%20millenia.

Svanberg, Ingvar, and Åsa Berggren. "Bumblebee Honey in the Nordic Countries." *Ethnobiology Letters* 9 no. 2 (2018): 312–318. https://doi.org/10.14237/ebl.9.2.2018.1383.

Swenson, Haylie. "The Political Insect: Bees as an Early Modern Metaphor for Human Hierarchy." *Shakespeare & Beyond.* (June 23, 2020). Folger Shakespeare Library. Retrieved April 6, 2022, from https://shakespeareandbeyond.folger.edu/2020/06/23/bees-metaphor-politics-hierarchy-wild-things/.

von Frisch, Karl., and Martin Lindauer. "The 'Language' and Orientation of the Honey Bee." *Annual Review of Entomology* 1 (1956): 45–58. https://doi.org/10.1146/annurev.en.01.010156.000401.

Williams, Paul. "The Distribution of Bumble Bee Colour Patterns Worldwide: Possible Significance for Thermoregulation, Crypsis, and

Warning Mimicry." *Biological Journal of the Linnean Society* 92 (2007): 97–118.

Wilson, Edward. O. *Biophilia*. Cambridge, MA: Harvard University Press, 1986.

Wilson, Joseph S., and Olivia Messinger Carril. *The Bees in Your Backyard: A Guide to North America's Bees*. Princeton, NJ: Princeton University Press, 2016.

Wilson-Rich, Noah. *The Bee: A Natural History*. Princeton, NJ: Princeton University Press, 2014.

Variations in bumble bee lifecycles:

Bumblebee Conservation Trust. "Winter Active Bumblebees." February 11, 2022. Retrieved April 1, 2023, from https://www.bumblebee conservation.org/winter-active-bumblebees/.

Cox, Darryl. "Bumblebees of the World Blog Series: #2 Bombus transversalis." *Bumblebee Conservation Trust*, February 14, 2019. Retrieved April 1, 2023, from https://www.bumblebeeconservation.org/bumblebees-of-the-world-blog-series-2-bombus-transversalis/.

Stelzer, Ralph J., Lars Chittka, Marc Carlton, and Thomas C. Ings. "Winter Active Bumblebees (*Bombus terrestris*) Achieve High Foraging Rates in Urban Britain." *PLoS ONE* 5, no. 3 (2010): e9559. https://doi.org/10.1371/journal.pone.0009559.

Taylor, Olivia Mariko, and Sydney A. Cameron. "Nest Construction and Architecture of the Amazonian Bumble Bee (Hymenoptera: Apidae)." *Apidologie* 34 (2003): 321–331. https://doi.org/10.1051/apido:2003035.

Gradual discovery of insect pollination over time (brief summary of the history of pollination):

Abrol, Dharam P. "Chapter 2: Historical Perspective." In *Pollination Biology: Biodiversity Conservation and Agricultural Production*, 25–35. Dordrecht: Springer Science+Business Media B.V., 2012. https://doi.org/10.1007/978-94-007-1942-2_2.

Early honey extraction methods, colonizers' hives, invention of Langstroth hives and other beekeeping equipment:

Oertel, Everett. "History of Beekeeping in the United States." *Beekeeping in the United States: Agriculture Handbook No. 335*, U.S. Department of Agriculture, January 1, 1967, pages 2–9. Retrieved November 28, 2023, from https://www.ars.usda.gov/ARSUserFiles/64133000/PDFFiles/1-100/093-Oertel--History%20of%20Beekeeping%20in%20the%20U.S..pdf.

Waite's experiments with pear orchards:

Waite, Merton B. "The Pollination of Pear Flowers." U.S. Department of Agriculture, Division of Vegetable Pathology, Bulletin No. 5 (1894). https://doi.org/10.5962/bhl.title.78597. Quote about startling results on page 3.

Quotes about insects, honey bees, and sweat bees on page 79.

Waite's pear orchard experiments might have been the idea behind moving beehives to orchards and crops for pollination:

Abrol, Dharam P. "Chapter 2: Historical Perspective." In *Pollination Biology: Biodiversity Conservation and Agricultural Production*, 25–35. Dordrecht: Springer Science+Business Media B.V., 2012. https://doi.org/10.1007/978-94-007-1942-2_2.

How commercial pollination developed alongside the evolution of transportation (including the quote from American Bee Journal):

Horn, Tammy. *Bees in America: How the Honey Bee Shaped a Nation.* Lexington, KY: The University of Kentucky Press, 2006. Pages 148–149.

Honey bees are still falling off trucks:

Brown, Desmond. "5 Million Bees Fall Off Truck on Guelph Line in Burlington, Ont." *CBC News*, August 30, 2023. Retrieved November 30, 2023, from https://www.cbc.ca/news/canada/hamilton/bees-off-truck-burlington-1.6951640.

Domestication of bumble bees and the birth and rise of the commercial bumble bee industry:

Velthuis, Hayo H. W., and Adriaan van Doorn. "A Century of Advances in Bumblebee Domestication and the Economic and Environmental Aspects of Its Commercialization for Pollination." *Apidologie* 37 (2006): 421–451. https://doi.org/10.1051/apido:2006019.

Chapter Three

Canadian Honey Council statistics on number of commercial hives and beekeepers:

Canadian Honey Council. "Industry Overview: Canadian Apiculture Industry." Retrieved November 30, 2023, from https://honeycouncil.ca/industry-overview/.

Production value of honey in Canada:

Agriculture and Agri-Food Canada. "Statistical Overview of the Canadian Honey and Bee Industry." August 2023. Retrieved November 30, 2023, from https://agriculture.canada.ca/en/sector/horticulture/reports/statistical-overview-canadian-honey-and-bee-industry-2022.

Number of managed honey bee colonies in the United States:

National Agricultural Statistics Service, United States Department of Agriculture. "Honey Bee Colonies." August 1, 2022. Retrieved November 30, 2023, from https://downloads.usda.library.cornell.edu/usda-esmis/files/rn301137d/kh04fx05c/qb98nn582/hcny0822.pdf.

Value of California almond crop and number of transported honey bee colonies for its pollination:

Economic Research Service, United States Department of Agriculture. "Thousands of Commercial Honey Bee Colonies Are Transported Long

Distances to Pollinate California Almonds." August 9, 2023. Retrieved November 30, 2023, from https://www.ers.usda.gov/data-products/chart-gallery/gallery/chart-detail/?chartId=107088.

Description of working for a large commercial honey bee business:
Anonymous. Personal interview, November 29, 2022.

David Hackenberg's cross-country pollination trips and crop value estimates:
Hackenberg Apiaries. "Pollination." Retrieved January 2, 2023, from https://hackenbergapiaries.us/pollination/.

David Hackenberg's discovery of his empty hives:
Cox-Foster, Diana, and Dennis vanEnglesdorp, "Solving the Mystery of the Vanishing Bees." *Scientific American*, April 1, 2009. Retrieved January 3, 2023, from https://www.scientificamerican.com/article/saving-the-honeybee/.
Portus, Rosamund. "An Ecological Whodunit: The Story of Colony Collapse Disorder." *Society & Animals* (2020): 1–19. https://doi.org/10.1163/15685306-BJA10026.

Jerry Hayes's account of the discovery of Colony Collapse Disorder:
Hayes, Jerry, December 9, 2022. Personal interview.

Newspaper headlines reporting disappearing honey bees:
Barrionuevo, Alexei. "Bees Vanish, and Scientists Race for Reasons." *The New York Times*. April 24, 2007. https://www.nytimes.com/2007/04/24/science/24bees.html (accessed December 4, 2022).
Barrionuevo, Alexei. "Honeybee Vanish, Leaving Keepers in Peril." *The New York Times*. February 27, 2007. https://www.nytimes.com/2007/02/27/business/27bees.html (accessed February 13, 2023).

Potter, Ned. "Honeybees Dying: Scientists Wonder Why, and Worry about Food Supply." *ABC News*. March 24, 2010. https://abcnews.go.com/Technology/honey-bees-dying-scientists-suspect-pesticides-disease-worry/story?id=10191391 (accessed February 13, 2023).

Semeniuk, Ivan. "News Review 2007: Where Have All the Honeybees Gone?" New Scientist. December 18, 2007. https://www.newscientist.com/article/mg19626355-400-news-review-2007-where-have-all-the-honeybees-gone/ (accessed February 13, 2023).

Zabarenko, Deborah. "Vanishing Honeybees Mystify Scientists." Reuters. April 22, 2007. https://www.reuters.com/article/us-bees-idUSN1930946620070422 (accessed February 13, 2023).

Agatha Christie quote:

Barrionnuevo, Alexei. "Honeybee Vanish, Leaving Keepers in Peril." *The New York Times*. February 27, 2007. https://www.nytimes.com/2007/02/27/business/27bees.html (accessed February 13, 2023).

List of theories to explain colony losses:

Barronuevo, Alexei. "Making a Beeline to Catch a Killer." Edmonton Journal. April 29, 2007. https://www.proquest.com/canadiannews/docview/253403165/85BB5709E8144D49PQ/148?accountid=48711 (accessed February 13, 2023).

Robbin Thorp quotes:

Sutter, John D. "The Old Man and the Bee." *CNN*. December 13, 2016. https://www.cnn.com/2016/12/11/us/vanishing-sutter-franklins-bumblebee/index.html (accessed February 21, 2023).

Robbin Thorp and Franklin's bumble bee:

The Xerces Society for Invertebrate Conservation, and Dr. Robbin W. Thorp. "Petition to List Franklin's Bumble Bee *Bombus franklini*

(Frison), 1921 as an Endangered Species under the U.S. Endangered Species Act." June 23, 2010. https://www.xerces.org/publications/policy-statements/petition-to-list-franklins-bumble-bee-under-esa-2010 (accessed February 21, 2023).

Sheila Colla's surveys of bumble bee species:
Colla, Sheila R., and Laurence Packer. "Evidence for Decline in Eastern North American Bumblebees (Hymenoptera: Apidae), with Special Focus on *Bombus affinis* Cresson." *Biodiversity and Conservation* 17 (2008): 1379–1391. https://doi.org/10.1007/s10531-008-9340-5.

Sheila Colla's quote about finding the rusty-patched bumble bee:
Colla, Sheila R., April 30, 2019. Personal interview.

Hawaii's yellow-faced bees:
Cole, Evan. "Yellow-Faced Bee—Hylaeus." U.S. Forest Service, United States Department of Agriculture. (No date.) https://www.fs.usda.gov/wildflowers/pollinators/pollinator-of-the-month/yellow-faced-bee.shtml (accessed February 27, 2023).
Magnacca, Karl N. "Conservation Status of the Endemic Bees of Hawai'i, *Hylaeus* (*Nesoprosopis*) Hymenoptera: Colletidae)." *Pacific Science* 61, no. 2 (2007): 173–190.

Reports of global decline of native bees before CCD:
Buchmann, Stephen, and John S. Ascher. "The Plight of Pollinating Bees." *Bee World* 86, no. 3 (2005): 71–74. https://doi.org/10.1080/0005772X.2005.11417316.
Goulson, Dave, Gillian. C. Lye, and Ben. Darvill. "Decline and Conservation of Bumble Bees." *Annual Review of Entomology* 53 (2008): 191–208. https://doi.org/10.1146/annurev.ento.53.103106.093454.

Concern for honey bee health before CCD:

Watanabe, Myrna E. "Pollination Worries Rise as Honey Bees Decline." *Science* 265, no. 5176 (1994): 1170. https://www.jstor.org/stable/2884859.

Chapter Four

Quotes from Jerry Hayes:

Hayes, Jerry, December 9, 2022. Personal interview.

Description of American and European Foulbrood:

Bee Research Laboratory: Beltsville, MD. "American Foulbrood Disease." USDA Agricultural Research Service, U.S. Department of Agriculture. (2016). https://www.ars.usda.gov/northeast-area/beltsville-md-barc/beltsville-agricultural-research-center/bee-research-laboratory/docs/american-foulbrood-disease/ (accessed February 28, 2023).

Texas Apiary Inspection Service (TAIS). "European Foulbrood." Texas A&M AgriLife Research. (No date.) https://txbeeinspection.tamu.edu/european-foulbrood/#:~:text=Like%20American%20Foulbrood%2C%20EFB%20targets,for%20both%20bees%20and%20beekeepers.&text=The%20causative%20agent%20of%20European,non%2Dspore%2Dforming%20bacterium (accessed February 28, 2023).

Honey bee problems before CCD:

Hayes, Jerry, December 9, 2022. Personal interview.

Watanabe, Myrna E. "Pollination Worries Rise as Honey Bees Decline." *Science* 265, no. 5176 (1994): 1170. https://www.jstor.org/stable/2884859.

Description of tracheal mites:

Bee Research Laboratory: Beltsville, MD. "Tracheal Mite." USDA Agricultural Research Service, U.S. Department of Agriculture. (2016). https://www.

ars.usda.gov/northeast-area/beltsville-md-barc/beltsville-agricultural-research-center/bee-research-laboratory/docs/tracheal-mite/ (accessed March 1, 2023).

Texas Apiary Inspection Service (TAIS). "Tracheal Mites." Texas A&M AgriLife Research. (No date.) https://txbeeinspection.tamu.edu/tracheal-mites/ (accessed March 1, 2023).

Watanabe, Myrna E. "Pollination Worries Rise as Honey Bees Decline." *Science* 265, no. 5176 (1994): 1170. https://www.jstor.org/stable/2884859.

Description of varroa mites:

Bee Research Laboratory: Beltsville, MD. "Varroa destructor." USDA Agricultural Research Service, U.S. Department of Agriculture. (2017). https://www.ars.usda.gov/northeast-area/beltsville-md-barc/beltsville-agricultural-research-center/bee-research-laboratory/docs/varroa-destructor/ (accessed March 1, 2023).

Texas Apiary Inspection Service (TAIS). "Varroa Mites." Texas A&M AgriLife Research. (No date.) https://txbeeinspection.tamu.edu/varroa-mites/ (accessed March 1, 2023).

Watanabe, Myrna E. "Pollination Worries Rise as Honey Bees Decline." *Science* 265, no. 5176 (1994): 1170. https://www.jstor.org/stable/2884859.

Comparison of size of varroa mites and quote about varroa mites:

University of Florida Honey Bee Research and Extension Lab. "A Video Field Guide to Beekeeping: Episode 4: Varroa Mites." Length: 25:28. Quote: 0:12 by Dr. Jamie Ellis. https://www.youtube.com/watch?v=S5vVrAy6CEU (accessed March 2, 2023).

Large-scale honey bee losses in the past:

Oldroyd, Benjamin P. "What's Killing American Honey Bees?" *PLoS Biology* 5, no. 6 (2007): e168. https://doi.org/10.1371/journal.pbio.0050168.

Underwood, Robyn M., and Dennis vanEngelsdorp. "Colony Collapse Disorder: Have We Seen This Before?" *Bee Culture Magazine* 135, no.7 (2007): 13–15.

Decline of managed honey bee colonies and the survey of honey bee colony losses across the United States, 2007–2008:

vanEngelsdorp, Dennis, Jerry Hayes Jr., Robyn M. Underwood, and Jeffery Pettis. "A Survey of Honey Bee Colony Losses in the U.S., Fall 2007 to Spring 2008." *PLoS ONE* 3, no. 12 (2008): e4071. https://doi.org/10.1371/journal.pone.0004071.

Monetary value of commercial honey bees in the United States:

Johnson, Renée. *Honey Bee Colony Collapse Disorder: Congressional Research Service Report for Congress.* (January 7, 2010). https://sgp.fas.org/crs/misc/RL33938.pdf (accessed March 4, 2023).

Response of the US government to CCD:

Johnson, Renée. *Honey Bee Colony Collapse Disorder: Congressional Research Service Report for Congress.* (January 7, 2010). https://sgp.fas.org/crs/misc/RL33938.pdf (accessed March 4, 2023).

United States Environmental Protection Agency. "Colony Collapse Disorder." (October 26, 2022). https://www.epa.gov/pollinator-protection/colony-collapse-disorder (accessed March 4, 2023).

Bee Research Laboratory, Beltsville, MD:

United States Department of Agriculture. "Bee Research Laboratory: Beltsville, MD." (No date.) https://www.ars.usda.gov/northeast-area/beltsville-md-barc/beltsville-agricultural-research-center/bee-research-laboratory/ (accessed March 4, 2023).

Possible causes of CDD:

Johnson, Renée. *Honey Bee Colony Collapse Disorder: Congressional Research Service Report for Congress*. (January 7, 2010). https://sgp.fas.org/crs/misc/RL33938.pdf (accessed March 4, 2023).

Sánchez-Bayo, Dave Goulson, Francesco Pennacchio, Francesco Nazzi, Koichi Goka, and Nicholas Desneux. "Are Bees Diseases Linked to Pesticides? A Brief Review." *Environment International* 89–90 (2016): 7–11. https://doi.org/10.1016/j.envint.2016.01.009.

Franklin's bumble bee decline:

The Xerces Society for Invertebrate Conservation, and Dr. Robbin W. Thorp. "Petition to List Franklin's Bumble Bee *Bombus franklini* (Frison), 1921 as an Endangered Species under the U.S. Endangered Species Act." (June 23, 2010). https://www.xerces.org/publications/policy-statements/petition-to-list-franklins-bumble-bee-under-esa-2010 (accessed February 21, 2023).

Buzz pollination:

Cooley, Hazel, and Mario Vallejo-Marín. "Buzz-Pollinated Crops: A Global Review and Meta-analysis of the Effects of Supplemental Bee Pollination in Tomato." *Journal of Economic Entomology* 114 no. 2 (2021): 505–519. https://doi.org/10.1093/jee/toab009.

Spread of *Nosema bombi* from commercial bumble bees to wild bumble bees:

The Xerces Society for Invertebrate Conservation, and Dr. Robbin W. Thorp. "Petition to List Franklin's Bumble Bee *Bombus franklini* (Frison), 1921 as an Endangered Species under the U.S. Endangered Species Act." (June 23, 2010). https://www.xerces.org/publications/policy-statements/petition-to-list-franklins-bumble-bee-under-esa-2010 (accessed February 21, 2023).

Velthuis, Hayo H. W., and Adriaan van Doorn. "A Century of Advances in Bumblebee Domestication and the Economic and Environmental Aspects of Its Commercialization for Pollination." *Apidologie* 37 (2006): 421–451. https://doi.org/10.1051/apido:2006019.

Rusty-patched bumble bee decline:

Cameron, Sydney A., Jeffrey D. Lozier, James P. Strange, Jonathan B Koch, Nils Cordes, Leellen F. Solter, and Terry L. Griswold. "Patterns of Widespread Decline in North American Bumble Bees." *Proceedings of the National Academy of Sciences* 108 no. 2 (2011): 662–667. https://doi.org/10.1073/pnas.1014743108.

Colla, Sheila R., Fawziah Gadallah, Leif Richardson, David Wagner, and Lawrence Gall. "Assessing Declines of North American Bumble Bees (Bombus spp.) Using Museum Specimens." *Biodiversity and Conservation* 21 (2012): 3585–3595. https://doi.org/10.1007/s10531-012-0383-2.

The Xerces Society for Invertebrate Conservation. "Petition to List the Rusty Patched Bumble Bee *Bombus affinis* (Cresson), 1863 as an Endangered Species under the U.S. Endangered Species Act." (January 31, 2013). https://www.xerces.org/publications/petitions-comments/petition-to-list-rusty-patched-bumble-bee-bombus-affinis-cresson (accessed March 6, 2023).

Experiment that examined pollen loads of commercial bumble bees in tomato greenhouses:

Whittington, Robin, Mark L. Winston, Chris Tucker, and Amy L. Parachnowitsch. "Plant-Species Identity of Pollen Collected by Bumblebees Placed in Greenhouses for Tomato Pollination." *Canadian Journal of Plant Science* 84, no. 2 (2004): 599–602.

Drifting bumble bees between wild and commercial colonies as a way to spread disease:

Hicks, Barry J., Brettney L. Pilgrim, Elizabeth Perry, and Heather D. Marshall. "Observations of Native Bumble Bees Inside of Commercial Colonies of *Bombus impatiens* (Hymenoptera: Apidae) and the Potential for Pathogen Spillover." *The Canadian Entomologist* 150 (2018): 520–531. https://doi.org/10.4039/tce.2018.28.

Pathogen spillover from commercial bumble bee colonies in greenhouses to wild bumble bee populations:

Colla, Sheila R., Michael C. Otterstatter, Robert J. Gegear, and James D. Thomson. "Plight of the Bumble Bee: Pathogen Spillover from Commercial to Wild Populations." *Biological Conservation* 129 (2006): 461–467. https://doi.org/10.1016/j.biocon.2005.11.013.

Hawaiian yellow-faced bees:

Graham, Jason R., Joshua W. Campbell, Sheldon Plentovich, and Cynthia B. A. King. "Nest Architecture of an Endangered Hawaiian Yellow-Faced Bee, *Hylaeus anthracinus* (Hymenoptera: Colletidae) and Potential Nest-Site Competition from Three Introduced Solitary Bees." *Pacific Science* 75, no. 3 (2021): 361–370.

Magnacca, Karl N. "Conservation Status of the Endemic Bees of Hawai'i, *Hylaeus* (*Nesoprosopis*) (Hymenoptera: Colletidae)." *Pacific Science* 61, no. 2 (2007): 173–190.

Magnacca, Karl N., and Bryan N. Danforth. "Evolution and Biogeography of Native Hawaiian *Hylaeus* bees (Hymenoptera: Colletidae)." *Cladistics* 22 (2006): 393–411.

Plentovich, Sheldon, Jason R. Graham, William P. Haines, and Cynthia B. A. King. "Invasive Ants Reduce Nesting Success of an Endangered Hawaiian Yellow-Faced Bee, *Hylaeus anthracinus*." *NeoBiota* 64 (2021): 137–154. https://doi.org/10.3897/neobiota.64.58670.

U.S. Fish and Wildlife Service: Pacific Islands. "Chat with a Scientist: Hawaiian Pollinator Edition." *Medium*. (December 29, 2021). https://medium.com/usfwspacificislands/chat-with-a-scientist-hawaiian-pollinator-edition-bb3004ca6b72 (accessed February 23, 2023).

Reference to Lars Chittka and Stephen Buchmann's books:
Buchmann, Stephen. *What a Bee Knows: Exploring the Thoughts, Memories, and Personalities of Bees*. Washington: Island Press, 2023.
Chittka, Lars. *The Mind of a Bee*. New Jersey: Princeton University Press, 2022.

Chapter Five

Wild bees visiting melon, watermelon, and almond flowers in Spain:
Rodrigo Gómez, Sara, Concepción Ornosa, Jaime García Gila, Javier Blasco-Aróstegui, Jesús Selfa, Miguel Guara, and Carlo Polidori. "Bees and Crops in Spain: An Update for Melon, Watermelon and Almond." *Annales de la Société Entomologique de France* 57, no. 1 (2021): 12–28. https://doi.org/10.1080/00379271.2020.1847191.

Large study by Rondeau, Willis Chan, and Pindar that identified wild bee visitors of major North American crops:
Rondeau, Sabrina, D. Susan Willis Chan, and Alana Pindar. "Identifying Wild Bee Visitors of Major Crops in North America with Notes on Potential Threats from Agricultural Practices." *Frontiers in Sustainable Food Systems* 6 (2022): 943237. https://doi.org/10.3389/fsufs.2022.943237.

Study of wild bee crop flower visitation in Virginia:
Adamson, Nancy L., T'ai H. Roulston, Rick D. Fell, and Donald E. Mullins. "From April to August—Wild Bees Pollinating Crops through the Growing Season in Virginia, USA." *Environmental Entomology* 41 no. 4 (2012): 813–821.

Wild bee pollinators of blueberry and cranberry in Newfoundland:
Hicks, Barry J. "Pollination of Lowbush Blueberry (*Vaccinium angustifolium*) in Newfoundland by Native and Introduced Bees." *Journal of the Acadian Entomological Society* 7 (2011): 108–118. https://acadianes.org/journal/papers/hicks_11-11.pdf.
Hicks, Barry J., and Julie Sircom. "Pollination of Commercial Cranberry (*Vaccinium macrocarpon* Ait.) by Native and Introduced Managed Bees in Newfoundland." *Journal of the Acadian Entomological Society* 12 (2016): 22–30. https://acadianes.org/journal/papers/hicks_16-2.pdf.

Wild pollinator communities in sweet cherry orchards in Belgium:
Eeraerts, Maxime, Guy Smagghe, and Ivan Meeus. "Pollinator Diversity, Floral Resources and Semi-Natural Habitat, Instead of Honey Bees and Intensive Agriculture, Enhance Pollination Service to Sweet Cherry." *Agriculture, Ecosystems and Environment* 284 (2019): 106586. https://doi.org/10.1016/j.agee.2019.106586.

Strawberry pollination in Québec:
MacInnis, Gail, and Jessica R. K. Forrest. "Pollination by Wild Bees Yields Larger Strawberries than Pollination by Honey Bees." *Journal of Applied Ecology* 56 (2019): 824–832. https://doi.org/10.1111/1365-2664.13344.

Wild bee pollination of apple orchards in Argentina (experimental manipulation):
Pérez-Méndez, Néstor, Georg K. S. Andersson, Fabrice Requier, Juliana Hipólito, Marcelo A. Aizen, Carolina L. Morales, Nancy García, Gerardo P. Gennari, and Lucas A. Garibaldi. "The Economic Cost of Losing Native Pollinator Species for Orchard Production." *Journal of Applied Ecology* 57 (2020): 599–608. https://doi.org/10.1111/1365-2664.13561.

Stats on global apple production and value:

Olhnuud, Aruhan, Yunhui Liu, David Makowski, Teja Tscharntke, Catrin Westphal, Panlong Wu, Meina Wang, and Wopke van der Werf. "Pollination Deficits and Contributions of Pollinators in Apple Production: A Global Meta-Analysis." *Journal of Applied Ecology* 59 (2022): 2911–2921. https://doi.org/10.1111/1365-2664.14279.

Weekers, Timothy, Leon Marshall, Nicolas Leclercq, Thomas J. Wood, Diego Cejas, Bianca Drepper, Michael Garratt, Louise Hutchinson, Stuart Roberts, Jordi Bosch, Laura Roquer-Beni, Patrick Lhomme, Denis Michez, Jean-Marc Molenberg, Guy Smagghe, Peter Vandamme, and Nicholas J. Vereecken. "Ecological, Environmental, and Management Data Indicate Apple Production Is Driven by Wild Bee Diversity and Management Practices." *Ecological Indicators* 139 (2022): 108880. https://doi.org/10.1016/j.ecolind.2022.108880.

Study of wild bee diversity and management practices in Western Europe and Morocco:

Weekers, Timothy, Leon Marshall, Nicolas Leclercq, Thomas J. Wood, Diego Cejas, Bianca Drepper, Michael Garratt, Louise Hutchinson, Stuart Roberts, Jordi Bosch, Laura Roquer-Beni, Patrick Lhomme, Denis Michez, Jean-Marc Molenberg, Guy Smagghe, Peter Vandamme, and Nicholas J. Vereecken. "Ecological, Environmental, and Management Data Indicate Apple Production Is Driven by Wild Bee Diversity and Management Practices." *Ecological Indicators* 139 (2022): 108880. https://doi.org/10.1016/j.ecolind.2022.108880.

Global meta-analysis of pollinators of apple crops:

Olhnuud, Aruhan, Yunhui Liu, David Makowski, Teja Tscharntke, Catrin Westphal, Panlong Wu, Meina Wang, and Wopke van der Werf. "Pollination Deficits and Contributions of Pollinators in Apple Production: A Global Meta-Analysis." *Journal of Applied Ecology* 59 (2022): 2911–2921. https://doi.org/10.1111/1365-2664.14279.

Halictus ligatus:

Boomsma, Jacobus. J., and George. C. Eickwort. "Colony Structure, Provisioning, and Sex Allocation in the Sweat Bee Halictus ligatus (Hymenoptera: Halictidae)." *Biological Journal of the Linnean Society* 48 (1993): 355–377. https://academic.oup.com/biolinnean/article/48/4/355/2661419.

Michener, Charles D., and Fred D. Bennett. "Geographical Variation in Nesting Biology and Social Organization of Halictus ligatus." *The University of Kansas Science Bulletin* 51, no. 7 (1977): 233–260. https://digitalcommons.usu.edu/bee_lab_mi/94.

Packer, Laurence. "The Social Organisation of Halictus ligatus (Hymenoptera; Halictidae) in Southern Ontario." *Canadian Journal of Zoology* 64 (1986): 2317–2324.

Rehan, Sandra. Personal interview. July 6, 2023.

Rehan, Sandra. M., Amanda. Rotella, Thomas. M. Onuferko, and Miriam. H. Richards. "Colony Disturbance and Solitary Nest Initiation by Workers in the Obligately Eusocial Sweat Bee, Halictus ligatus." *Insectes Sociaux* 60 (2013): 389–392. https://doi.org/10.1007/s00040-013-0304-8.

Squash bees (general information):

Brochu, Kristen. K., Shelby. J. Fleischer, and Margarita. M. López-Uribe. "Biology of the Squash Bee, *Eucera* (*Peponapis*) *pruinosa*." (2021). Penn State Extension (Booklet). https://lopezuribelab.com/squash-bee-biology/.

Willis Chan, D. Susan. Email communications April 17 and April 18, 2023.

Willis Chan, D. Susan, and Nigel E. Raine. "Hoary Squash Bees (*Eucera pruinosa*: Hymenoptera: Apidae) Provide Abundant and Reliable Pollination Services to Cucurbita Crops in Ontario (Canada)." *Environmental Entomology* 50, no. 4 (2021): 968–981. https://doi.org/10.1093/ee/nvab045.

Bumble bees and honey bees rejecting squash pollen:

Brochu, Kristen K., Maria T. van Dyke, Nelson J. Milano, Jessica D. Petersen, Scott H. McArt, Brian A. Nault, André Kessler, and Bryan N. Danforth. "Pollen Defenses Negatively Impact Foraging and Fitness in a Generalist Bee (*Bombus impatiens*: Apidae)." *Scientific Reports* 10, no. 3112 (2020): 1–12. https://doi.org/10.1038/s41598-020-58274-2.

López-Uribe, Margarita. "The Ecology and Evolution of Squash Bees and How Humans Have Influenced Their Recent History." Bee Biogeography and Systematics Talks (BeeBST): An International Webinar Series. April 26, 2023. https://www.yorku.ca/bees/packer/ (not recorded).

Seasonal and daily synchronicity between squash bees and cucurbit plants:

Brochu, Kristen. K., Shelby. J. Fleischer, and Margarita. M. López-Uribe. "Biology of the Squash Bee, *Eucera (Peponapis) pruinosa*." (2021). Penn State Extension (Booklet). https://lopezuribelab.com/squash-bee-biology/.

Willis Chan, D. Susan, and Nigel E. Raine. "Phenological Synchrony between the Hoary Squash Bee (*Eucera pruinosa*) and Cultivated Acorn Squash (*Cucurbita pepo*) Flowering Is Imperfect at a Northern Site." *Current Research in Insect Science* 1 (2021): 100022. https://doi.org/10.1016/j.cris.2021.100022.

Squash bees expanded their geographical range with the domestication of crops:

López-Uribe, Margarita. "The Ecology and Evolution of Squash Bees and How Humans Have Influenced Their Recent History." Bee Biogeography and Systematics Talks (BeeBST): An International Webinar Series. April 26, 2023. https://www.yorku.ca/bees/packer/ (not recorded).

López-Uribe, Margarita M., James H. Crane, Robert L. Minckley, and Bryan N. Danforth. "Crop Domestication Facilitated Rapid Geographical

Expansion of a Specialist Pollinator, the Squash Bee *Peponapis pruinosa.*" *Proceedings of the Royal Society B* 283 (2016): 20160443. http://dx.doi.org/10.1098/rspb.2016.0443.

Pope, Nathaniel S., Avehi Singh, Anna K. Childers, Karen M. Kapheim, Jay D. Evans, and Margarita M. López-Uribe. "The Expansion of Agriculture Has Shaped the Recent Evolutionary History of a Specialized Squash Pollinator." *Proceedings of the National Academy of Sciences (PNAS)* 120, no. 15 (2023): e2208116120. https://doi.org/10.1073/pnas.2208116120.

Quote about domesticated pumpkin and squash smelling different from wild buffalo gourd:
Pope et al. (2023), p. 6.

Effects of pesticides on squash bees:
Willis Chan, D. Susan, and Nigel E. Raine. "Population Decline in a Ground-Nesting Solitary Squash Bee (Eucera pruinosa) Following Exposure to a Neonicotinoid Insecticide Treated Crop (Cucurbita pepo)." *Scientific Reports* 11 (2021): 4241. https://doi.org/10.1038/s41598-021-83341-7.

Quote from Sue Willis Chan about her research with pesticides:
Willis Chan, Susan. Personal communication. December 1, 2022.

Pollinator-dependent crops outpacing the supply of pollinators:
Aizen, Marcelo A., and Lawrence D. Harder. "The Global Stock of Domesticated Honey Bees Is Growing Slower than Agricultural Demand for Pollination." *Current Biology* 19 (2009): 915–918. https://doi.org/10.1016/j.cub.2009.03.071.

U.S. pumpkin industry statistics:
USDA Economic Research Service. "Pumpkins: Background & Statistics." May 24, 2023. Retrieved May 27, 2023 from https://www.ers.usda.gov/newsroom/trending-topics/pumpkins-background-statistics/.

Wild bees pollinating pumpkin crops:
McGrady, Carley. M., Rachael. Troyer, and Shelby. J. Fleischer. "Wild Bee Visitation Rates Exceed Pollination Thresholds in Commercial Cucurbita Agroecosystems." *Journal of Economic Entomology* 113 no. 2 (2020): 562–574. https://doi.org/10.1093/jee/toz295.

Quote from Sue Willis Chan about squash bees out-competing honey bees:
Willis Chan, Susan. Personal interview. December 1, 2022.

Quote from Margarita López-Uribe about the expansion of the squash bee population:
Environment News Service. "Squash Bee Populations Growing Fast While Other Bees Fail." April 4, 2023. Retrieved May 31, 2023, from https://ens-newswire.com/squash-bee-populations-growing-fast-while-other-bees-fail/.

High levels and diversity of parasites found in squash bees:
Jones, Laura J., Avehi Singh, Rudolf J. Schilder, and Margarita M. López-Uribe. "Squash Bees Host High Diversity and Prevalence of Parasites in the Northeastern United States." *Journal of Invertebrate Pathology* 195 (2022): 107848. https://doi.org/10.1016/j.jip.2022.107848.

Percentage of flowering species that are crop plants:
Ollerton, Jeff, Rachael Winfree, and Sam Tarrant. "How Many Flowering Plants Are Pollinated by Animals?" *Oikos* 120 (2011): 321–326. https://doi.org/10.1111/j.1600-0706.2010.18644.x.

Crop pollination is achieved by a small number of wild bee species and is thus too narrow a lens through which to view pollinator conservation:
Kleijn, David, Rachael Winfree, Ignasi Bartomeus, Luísa G. Carvalheiro, Mickaël Henry, Rufus Isaacs, et al. "Delivery of Crop Pollination Services

Is an Insufficient Argument for Wild Pollinator Conservation. *Nature Communications* 6 (2015): 7414. https://doi.org/10.1038/ncomms8414.

Chapter Six

Description of *Bombus dahlbomii*:

Cox, Darryl. "Bumblebees of the World...#1 Bombus dahlbomii." Bumblebee Conservation Trust, January 16, 2019. Retrieved August 28, 2023, from https://www.bumblebeeconservation.org/bumblebees-of-the-world-1-bombus-dahlbomii/.

Morales, Carolina L., Marina P. Arbetman, Sydney A. Cameron, and Marcelo A. Aizen. "Rapid Ecological Replacement of a Native Bumble Bee by Invasive Species." *Frontiers in Ecology and the Environment* 11 no. 10 (2013): 529–534. https://doi.org/10.1890/120321.

Morales, Carolina, Jose Montalva, Marina P. Arbetman, Marcelo A. Aizen, Aline C. Martins, and Daniel Paiva Silva. "Does Climate Change Influence the Current and Future Projected Distribution of an Endangered Species? The Case of the Southernmost Bumblebee in the World." *Journal of Insect Conservation* 26 (2022): 257–269. https://doi.org/10.1007/s10841-022-00384-5.

Toth, Amy. Personal communication, April 15, 2019.

Quotes and experiences of José Montalva:

Benjamin, Alison. "The Battle to Save the World's Biggest Bumblebee from Extinction." *The Guardian.* (May 4, 2019). Retrieved September 21, 2023 from https://www.theguardian.com/environment/2019/may/04/the-battle-to-save-the-worlds-biggest-bumblebee-from-european-invaders.

Description of amancay and its relationship with *Bombus dahlbomii*:

Morales, Carolina L., Marina P. Arbetman, Sydney A. Cameron, and Marcelo A. Aizen. "Rapid Ecological Replacement of a Native Bumble

Bee by Invasive Species." *Frontiers in Ecology and the Environment* 11 no. 10 (2013): 529–534. https://doi.org/10.1890/120321.

Summary of importation of bumble bees into Chile and spread to Argentina with description of effects:

Aizen, Marcelo A., Cecilia Smith-Ramírez, Carolina L. Morales, Lorena Vieli, Agustín Sáez, Rodrigo M. Barahona-Segovia, Marina P. Arbetman, José Montalva, Lucas A. Garibaldi, David W. Inouye, and Lawrence D. Harder. "Coordinated Species Importation Policies Are Needed to Reduce Serious Invasions Globally: The Case of Alien Bumblebees in South America." *Journal of Applied Ecology* 56 (2019): 100–106. https://doi.org/10.1111/1365-2664.13121.

Bombus dahlbomii on the IUCN Red List of Threatened Species:

Morales, Carolina., Jose Montalva, Marina Arbetman, Marcelo A. Aizen, Cecilia Smith-Ramírez, Lorena Vieli, and Rich Hatfield. *Bombus dahlbomii. The IUCN Red List of Threatened Species* 2016: e.T21215142A100240441. Retrieved September 9, 2023, from https://www.iucnredlist.org/species/21215142/100240441.

Quotes by Marcelo Aizen on nectar robbing by Bombus terrestris:

Aizen, Marcelo A., September 29, 2023. Personal interview.

Nectar robbing and why it is harmful:

Goulson, Dave. "Effects of Introduced Bees on Native Ecosystems." *Annual Review of Ecology, Evolution, & Systematics* 34 (2003): 1–26. https://doi.org/10.1146/annurev.ecolsys.34.011802.132355.

Quotes by Amy Toth about Bombus dahlbomii as a secondary nectar robber:

Toth, Amy, October 6, 2023. Personal interview.

Nectar robbing of *Fuchsia* by *Bombus terrestris* and *Bombus dahlbomii* as a secondary robber:

Rosenberger, Nick M., Marcelo A. Aizen, Rachel G. Dickson, and Lawrence D. Harder. "Behavioural Responses by a Bumble Bee to Competition with a Niche-Constructing Congener." *Journal of Animal Ecology* 91 (2022): 580–592. https://doi.org/10.1111/1365-2656.13646.

Other effects of introduced bees (nesting sites, competition for flowers, pathogens):

Goulson, Dave. "Effects of Introduced Bees on Native Ecosystems." *Annual Review of Ecology, Evolution, & Systematics* 34 (2003): 1–26. https://doi.org/10.1146/annurev.ecolsys.34.011802.132355.

Quote from Cecilia Smith-Ramírez about colonizers spreading foreign disease to native peoples:

Benjamin, Alison. "The Battle to Save the World's Biggest Bumblebee from Extinction." *The Guardian.* (May 4, 2019). Retrieved September 21, 2023 from https://www.theguardian.com/environment/2019/may/04/the-battle-to-save-the-worlds-biggest-bumblebee-from-european-invaders.

Study showing infection of *Apicystis bombi* in *Bombus dahlbomii*, *Bombus ruderatus*, and *Bombus terrestris*:

Arbetman, Marina P., Ivan Meeus, Carolina L. Morales, Marcelo A. Aizen, and Guy Smagghe. "Alien Parasite Hitchhikes to Patagonia on Invasive Bumblebee." *Biological Invasions* 15 (2013): 489–494. https://doi.org/10.1007/s10530-012-0311-0.

Commercial *Bombus impatiens* and evidence of pathogen spillover from greenhouses to wild bumble bee populations:

Colla, Sheila R., Michael C. Otterstatter, Robert J. Gegear, and James D. Thomson. "Plight of the Bumble Bee: Pathogen Spillover from

Commercial to Wild Populations." *Biological Conservation* 129 (2006): 461–467. https://doi.org/10.1016/j.biocon.2005.11.013.

Pathogens found in commercial *Bombus terrestris* colonies in England:
Graystock, Peter, Kathryn Yates, Sophie E. F. Evison, Ben Darvill, Dave Goulson, and William O. H. Hughes. "The Trojan Hives: Pollinator Pathogens, Imported and Distributed in Bumblebee Colonies." *Journal of Applied Ecology* 50 (2013): 1207–1215. https://doi.org/10.1111/1365-2664.12134.

Pathogens found in commercial *Bombus impatiens* in Mexico:
Sachman-Ruiz, Bernardo, Verónica Narváez-Padilla, and Enrique Reynaud. "Commercial Bombus Impatiens as Reservoirs of Emerging Infectious Diseases in Central México." *Biological Invasions* 17 (2015): 2043–2053. https://doi.org/10.1007/s10530-015-0859-6.

Dave Goulson's statements of how little we know about bee pathogens and quote about how more research is urgently needed:
Goulson, Dave. "Effects of Introduced Bees on Native Ecosystems." *Annual Review of Ecology, Evolution, & Systematics* 34 (2003): 1–26. (Quote on page 13). https://doi.org/10.1146/annurev.ecolsys.34.011802.132355.

Quote from Dave Goulson about how imported bees are often cheaper than raising native bees:
Benjamin, Alison. "The Battle to Save the World's Biggest Bumblebee from Extinction." *The Guardian.* (May 4, 2019). Retrieved September 21, 2023 from https://www.theguardian.com/environment/2019/may/04/the-battle-to-save-the-worlds-biggest-bumblebee-from-european-invaders.

Research twenty years later on pathogens and commercial bumble bees, and the need for a clean stock certification program:

Figueroa, Laura L., Ben M. Sadd, Amber D. Tripodi, James P. Strange, Sheila R. Colla, Laurie Davies Adams, et al. "Endosymbionts That Threaten Commercially Raised and Wild Bumble Bees (*Bombus* spp.)." *Journal of Pollination Ecology* 33 no. 2 (2023): 14–36. https://doi.org/10.26786/1920-7603(2023)713.

Strange, James P., Sheila R. Colla, Laurie Davis Adams, Michelle A. Duennes, Elaine C. Evans, Laura L. Figueroa, et al. "An Evidence-Based Rationale for a North American Commercial Bumble Bee Clean Stock Certification Program." *Journal of Pollination Ecology* 33 no. 1 (2023): 1–13. https://doi.org/10.26786/1920-7603(2023)721.

Managed honey bees creating potential competition with native pollinators like invasive *Bombus terrestris* has done:

Goulson, Dave. "Effects of Introduced Bees on Native Ecosystems." *Annual Review of Ecology, Evolution, & Systematics* 34 (2003): 1–26. https://doi.org/10.1146/annurev.ecolsys.34.011802.132355.

Flower damage from overabundance of bees, with the example of raspberries:

Aizen, Marcelo A., September 29, 2023. Personal interview.

Sáez, A., Carolina L. Morales, Lorena Y. Ramos, and Marcelo A. Aizen. "Extremely Frequent Bee Visits Increase Pollen Deposition but Reduce Drupelet Set in Raspberry." *Journal of Applied Ecology* 51 (2014): 1603–1612.

Information about Salvemos Nuestro Abejorro:

Montalva, José. "Modeling the Distribution of Native and Invasive Species of Bumble Bees (Hymenoptera: Apidae) in Chile, Using Citizen Science Data." Masters Thesis, University of Oklahoma (2021).

Study on the effects of climate change on *Bombus dahlbomii* using citizen science data:

Morales, Carolina L., José Montalva, Marina P. Arbetman, Marcelo A. Aizen, Aline C. Martins, and Daniel Paiva Silva. "Does Climate Change Influence the Current and Future Projected Distribution of an Endangered Species? The Case of the Southernmost Bumblebee in the World." *Journal of Insect Conservation*, 26 (2022): 257–269.

Bombus dahlbomii and native Chilean Mapuche peoples:

Montalva, José. "Modeling the Distribution of Native and Invasive Species of Bumble Bees (Hymenoptera: Apidae) in Chile, Using Citizen Science Data." Masters Thesis, University of Oklahoma (2021).

Montalva, José, Leah S. Dudley, Joaquín E. Sepúlveda, and Cecilia Smith-Ramírez. "The Giant Bumble Bee (*Bombus dahlbomii*) in Mapuche Cosmovision." *Ethnoentomology*, 4 (2020): 1–11.

Quote about the loss of *Bombus dahlbomii* being an erosion of native South American culture:

Montalva, José, Leah S. Dudley, Joaquín E. Sepúlveda, and Cecilia Smith-Ramírez. "The Giant Bumble Bee (*Bombus dahlbomii*) in Mapuche Cosmovision." *Ethnoentomology*, 4 (2020): 1–11. Quote on page 5.

Scientists Montalva, Aizen, Morales, et al., working to inform policy:

Aizen, Marcelo A., Cecilia Smith-Ramírez, Carolina L. Morales, Lorena Vieli, Agustín Sáez, Rodrigo M. Barahona-Segovia, et al. "Coordinated Species Importation Policies Are Needed to Reduce Serious Invasions Globally: The Case of Alien Bumblebees in South America." *Journal of Applied Ecology* 56 (2019): 100–106.

Possible reasons why *Bombus terrestris* is so successful and why some bumble bee species are in decline:

Dafni, Amots, Peter Kevan, Caroline L. Gross, and Koichi Goka. "*Bombus terrestris*, Pollinator, Invasive and Pest: An Assessment of Problems Associated with Its Widespread Introductions for Commercial Purposes." *Applied Entomology and Zoology* 45 no. 1 (2010): 101–113. https://doi.org/10.1303/aez.2010.101.

Manfredini, Fabio, Marina Arbetman, and Amy L. Toth. "A Potential Role for Phenotypic Plasticity in Invasions and Declines of Social Insects." *Frontiers in Ecology and Evolution*, 7 (2019): 375. https://doi.org/10.3389/fevo.2019.00375.

Chapter Seven

Definition of invasive species:

Russo, Laura. "Positive and Negative Impacts of Non-Native Bee Species around the World." *Insects* 7 no. 4 (2016): 69–91. https://doi.org/10.3390/insects7040069.

Research on competition between native and non-native bee species is contradictory or inconclusive:

Russo, Laura. "Positive and Negative Impacts of Non-Native Bee Species around the World." *Insects* 7 no. 4 (2016): 69–91. https://doi.org/10.3390/insects7040069.

Why demonstrating the presence of competition is so challenging:

Goulson, Dave. "Effects of Introduced Bees on Native Ecosystems." *Annual Review of Ecology, Evolution, & Systematics* 34 (2003): 1–26. https://doi.org/10.1146/annurev.ecolsys.34.011802.132355.

Disruption of plant-pollinator networks near orange groves in Spain:
Magrach, Ainhoa, Juan P. González-Varo, Mathieu Boiffier, Montserrat Vilà, and Ignasi Bartomeus. "Honeybee Spillover Reshuffles Pollinator Diets and Affects Plant Reproductive Success." *Nature Ecology & Evolution* 1 (2017): 1299–1307. https://doi.org/10.1038/s41559-017-0249-9.

Managed honey bees disrupting plant-pollinator networks in the Canary Islands:
Valido, Alfredo, María C. Rodríguez-Rodríguez, and Pedro Jordano. "Honeybees Disrupt the Structure and Functionality of Plant-Pollinator Networks." *Scientific Reports* 9 (2019): 4711. https://doi.org/10.1038/s41598-019-41271-5. Quote from abstract.

Managed honey bees disrupting plant-pollinator networks in Sierra Nevada, California:
Page, Maureen, and Neal M. Williams. "Honey Bee Introductions Displace Native Bees and Decrease Pollination of a Native Wildflower." *Ecology* 104 (2023): e3939. https://doi.org/10.1002/ecy.3939.

Islands and protected areas being particularly vulnerable to species introductions, and a review of the effects of introduced managed honey bees (and commercial bumble bees) in general:
Geslin, Benoît, Benoit Gauzens, Mathilde Baude, Isabelle Dajoz, Colin Fontaine, et al. "Massively Introduced Managed Species and Their Consequences for Plant-Pollinator Interactions." *Advances in Ecological Research* 57 (2017): 147–199. https://doi.org/10.1016/bs.aecr.2016.10.007.

Spread of parasites and pathogens as an unequivocal negative impact of invasive species:
Russo, Laura. "Positive and Negative Impacts of Non-Native Bee Species around the World." *Insects* 7 no. 4 (2016): 69–91. https://doi.org/10.3390/insects7040069.

Spillover of deformed wing virus and black queen cell virus to wild bumble bees near apiaries infected with the viruses:

Alger, Samantha A., P. Alexander Burnham, Humberto F. Boncristiani, and Alison K. Brody. "RNA Virus Spillover from Managed Honeybees (*Apis mellifera*) to Wild Bumblebees (*Bombus* spp.)." *PLoS ONE* 14 no. 6 (2019): e0217822. https://doi.org/10.1371/journal.pone.0217822.

What DWV and BQCV does to bees and evidence that honey bees deposit viruses on flowers:

Alger, Samantha A., P. Alexander Burnham, and Alison K. Brody. "Flowers as Viral Hot Spots: Honey bees (Apis mellifera) Unevenly Deposit Viruses across Plant Species." *PLoS ONE* 14 no. 9 (2019): e0221800. https://doi.org/10.1371/journal.pone.0221800.

DWV transmitted from honey bees to bumble bees through shared flowers, and a plant-pollinator transmission model:

Burnham, Philip Alexander, Samantha A. Alger, Brendan Case, Humberto Boncristiani, Laurent Hébert-Dufresne, and Alison K. Brody. "Flowers as Dirty Doorknobs: Deformed Wing Virus Transmitted between *Apis mellifera* and *Bombus impatiens* through Shared Flowers." *Journal of Applied Ecology* 58 (2021): 2065–2074. https://doi.org/10.1111/1365-2664.13962.

Stingless bees and pathogen spillover from managed honey bees to *Tetragonula hockingsi* in Australia:

Purkiss, Terence, and Lori Lach. "Pathogen Spillover from *Apis mellifera* to a Stingless Bee." *Proceedings of the Royal Society B* 286 (2019): 20191071. https://doi.org/10.1098/rspb.2019.1071.

Rise of managed honey bees in the Mediterranean Basin:

Herrera, Carlos M. "Gradual Replacement of Wild Bees by Honey Bees in Flowers of the Mediterranean Basin over the Last 50 Years."

Proceedings of the Royal Society B 287 (2020): 20192657. https://doi. org/10.1098/rspb.2019.2657. Quote pages 5–6.

Chapter Eight

History of Alvéole:

Alvéole. "About Us." Retrieved November 18, 2023, from https://www. alveole.buzz/about-us/.

Helfenbaum, Wendy. "Generating Plenty of Buzz with Urban Beekeeping." *McGill News Alumni Magazine.* (2016). Retrieved January 28, 2022, from https://mcgillnews.mcgill.ca/s/1762/news/interior.aspx?sid=1762& gid=2&pgid=1361.

Number of beehives Alvéole has around the world:

Alvéole (Press Release). "Alvéole's Swarming into New Locations." December 6, 2021. Retrieved November 18, 2023, from https://www.alveole.buzz/ press-release-12-new-cities-for-alveole/.

Quotes throughout from Alvéole representatives I spoke with:

Interview, September 1, 2022.

Quote about Alvéole reconnecting people to nature:

Helfenbaum, Wendy. (see above)

Description of Alvéole's tracking of client's impact and biodiversity data:

Alvéole. (2023). "Environmental Impact Tracking: Track Your Building's Impact with Science-Based Biodiversity Data." Retrieved November 18, 2023, from https://www.alveole.buzz/environmental-reporting/.

Quote about the history of The Best Bees Company (and quotes throughout):
Interview, December 9, 2022.

Background and description of The Best Bees Company:
The Best Bees Company. "About Us." (2023). Retrieved November 18, 2023, from https://bestbees.com/about/.

The Best Bees Company. "Corporate Beekeeping Services & Pollinator Programs." (2023). Retrieved November 18, 2023, from https://bestbees.com/corporate-beekeeping-services/.

The Best Bees Company. "Residential Beekeeping." (2023). Retrieved November 18, 2023, from https://bestbees.com/residential-beekeeping-services/.

Interview, December 9, 2022.

Quote from sideliner beekeeper about urban beekeeping, "save the bees," and bee washing:
Anonymous. Interview, November 16, 2022.

The term *bee washing* and research on bee hotels and other products:
Alton, Karin, and Francis L. W. Ratnieks. "Caveat Emptor: Do Products Sold to Help Bees and Pollinating Insects Actually Work?" *Bee World* 97 no. 2 (2020): 57–60. https://doi.org/10.1080/0005772X.2019.1702271.

Colla, Sheila R. "The Potential Consequences of 'Bee Washing' on Wild Bee Health and Conservation." *International Journal for Parasitology: Parasites and Wildlife* 18 (2022): 30–32. https://doi.org/10.1016/j.ijppaw.2022.03.011.

MacIvor, J. Scott, and Laurence Packer. "'Bee Hotels' as Tools for Native Pollinator Conservation: A Premature Verdict?" *PLoS ONE* 10 no. 3 (2015): e0122126. https://doi.org/10.1371/journal.pone.0122126.

Unsustainability of urban beekeeping in Swiss cities:

Casanelles-Abella, Joan, and Marco Moretti. "Author Correction: Challenging the Sustainability of Urban Beekeeping Using Evidence from Swiss Cities." *npj Urban Sustainability* 2 no. 14 (2022). https://doi.org/10.1038/s42949-022-00059-9.

Casanelles-Abella, Joan, and Marco Moretti. "Challenging the Sustainability of Urban Beekeeping Using Evidence from Swiss Cities." *npj Urban Sustainability* 2 no. 3 (2022). https://doi.org/10.1038/s42949-021-00046-6.

Unsustainability of urban beekeeping in London, United Kingdom:

Stevenson, Philip C., Martin I. Bidartondo, Robert Blackhall-Miles, Timothy R. Cavagnaro, Amanda Cooper, Benoît Geslin, et al. "The State of the World's Urban Ecosystems: What Can We Learn from Trees, Fungi, and Bees?" *Plants People Planet* 2 (2020): 482–498. https://doi.org/10.1002/ppp3.10143.

Quote from Casanelles-Abella about their work being inspired by the London Beekeepers Association:

Casanelles, Abella, Joan. Email communication, March 25, 2024.

The number of extra plants needed to accommodate urban beehives:

Colla, Sheila R. Email communication, December 14, 2022.

Decline in wild bee diversity on the Island of Montréal with the increase in urban beekeeping:

MacInnis, Gail, Etienne Normandin, and Carly D. Ziter. "Decline in Wild Bee Species Richness Associated with Honey Bee (*Apis mellifera* L.) Abundance in an Urban Ecosystem." *PeerJ* 11 (2023): e14699. https://doi.org/10.7717/peerj.14699.

Decline in wild pollinator activity with urban beekeeping in Paris, France:

Ropars, Lise, Isabelle Dajoz, Colin Fontaine, Audrey Muratet, and Benoît Geslin. "Wild Pollinator Activity Negatively Related to Honey Bee Colony Densities in Urban Context." *PLoS ONE* 14 no. 9 (2019): e0222316. https://doi.org/10.1371/journal.pone.0222316.

Sheila Colla on the consequences of bee washing:

Colla, Sheila R. "The Potential Consequences of 'Bee Washing' on Wild Bee Health and Conservation." *International Journal for Parasitology: Parasites and Wildlife* 18 (2022): 30–32. https://doi.org/10.1016/j.ijppaw.2022.03.011.

Thomas Seeley's quote about cows and honey bees sharing the same fate:

Seeley, Thomas D. (2019). *The Lives of Bees: The Untold Story of the Honey Bee in the Wild.* Princeton, NJ: Princeton University Press. Quote page 86.

Honey bees as special cases of livestock:

Casanelles-Abella, Joan, and Marco Moretti. "Challenging the Sustainability of Urban Beekeeping Using Evidence from Swiss Cities." *npj Urban Sustainability* 2 no. 3 (2022). https://doi.org/10.1038/s42949-021-00046-6.

Quote from Sheila Colla about beekeeping companies being transparent about disease levels of their hives and pathogen spillover:

Colla, Sheila R. Email communication, December 14, 2022.

Sheila Colla quoting the precautionary principle:

Colla, Sheila R. "The Potential Consequences of 'Bee Washing' on Wild Bee Health and Conservation." *International Journal for Parasitology:*

Parasites and Wildlife 18 (2022): 30–32. https://doi.org/10.1016/j.ijppaw.2022.03.011. Quote on page 31.

Zach Portman's essay on keeping chickens to save the birds:

Portman, Zachary. (2022). "How I'm Helping to Save the Birds by Keeping Chickens." Retrieved February 3, 2022, from https://www.zportman.com/writing.html.

Conclusion

Beekeeper quote from a *New Yorker* article that references "When Harry Met Sally":

Knight, Sam. "Hive Mind: Is Beekeeping Wrong?" *The New Yorker*. (August 21, 2023). Retrieved December 7, 2023, from https://www.newyorker.com/magazine/2023/08/28/is-beekeeping-wrong.

Recommended books by Lorraine Johnson, Sheila Colla, and Ann Sanderson (illustrator):

Johnson, Lorraine, and Sheila Colla. *A Garden for the Rusty-Patched Bumblebee: Creating Habitat for Native Pollinators* (Ontario and Great Lakes Edition). Madiera Park, BC: Douglas & McIntyre, 2022.

Johnson, Lorraine, and Sheila Colla. *A Northern Gardener's Guide to Native Plants and Pollinators: Creating Habitat in the Northeast, Great Lakes, and Upper Midwest*. Washington, DC: Island Press, 2023.

Acknowledgements

M UCH GRATITUDE TO Sam Hiyate and Kathryn Willms at The Rights Factory, and to my incredible agent, Stacey Kondla.

My heartfelt thanks to my editors, Ken Whyte, Ian Coutts, and Shalomi Ranasinghe. Thank you to Leah Ciani and Jordan Lunn for making it look so beautiful, and to Serina Mercier for getting the word out.

I am forever grateful to Dr. Henry Lickers, who so kindly and generously shared his people's knowledge with me, so that I can share it with readers.

Thank you to all of the beekeepers who took the time to speak with me and answer my questions.

Enormous gratitude to the scientists featured in this book, and to those I didn't have room to include. Your work and efforts are so tremendously important.

And last but certainly not least, to my personal cheering squad, Cia and Stephen. Thank you deeply for your endless love, encouragement, and support. A++ to you both.